LONDON MATHEMATICAL SOCIETY LEC

Managing Editor: Professor M. Reid, Mathematics Institute,
University of Warwick, Coventry CV4 7AL, United Kingdom

The titles below are available from booksellers, or from Cambric

234 Introduction to subfactors, V. JONES & V.S. SUNDER
235 Number theory: Séminaire de théorie des nombres de Paris 1993–94, S. DAVID (ed)
236 The James forest, H. FETTER & B. GAMBOA DE BUEN
237 Sieve methods, exponential sums, and their applications in number theory, G.R.H. GREAVES *et al* (eds)
238 Representation theory and algebraic geometry, A. MARTSINKOVSKY & G. TODOROV (eds)
240 Stable groups, F.O. WAGNER
241 Surveys in combinatorics, 1997, R.A. BAILEY (ed)
242 Geometric Galois actions I, L. SCHNEPS & P. LOCHAK (eds)
243 Geometric Galois actions II, L. SCHNEPS & P. LOCHAK (eds)
244 Model theory of groups and automorphism groups, D.M. EVANS (ed)
245 Geometry, combinatorial designs and related structures, J.W.P. HIRSCHFELD *et al* (eds)
246 p-Automorphisms of finite p-groups, E.I. KHUKHRO
247 Analytic number theory, Y. MOTOHASHI (ed)
248 Tame topology and O-minimal structures, L. VAN DEN DRIES
249 The atlas of finite groups - Ten years on, R.T. CURTIS & R.A. WILSON (eds)
250 Characters and blocks of finite groups, G. NAVARRO
251 Gröbner bases and applications, B. BUCHBERGER & F. WINKLER (eds)
252 Geometry and cohomology in group theory, P.H. KROPHOLLER, G.A. NIBLO & R. STÖHR (eds)
253 The q-Schur algebra, S. DONKIN
254 Galois representations in arithmetic algebraic geometry, A.J. SCHOLL & R.L. TAYLOR (eds)
255 Symmetries and integrability of difference equations, P.A. CLARKSON & F.W. NIJHOFF (eds)
256 Aspects of Galois theory, H. VÖLKLEIN, J.G. THOMPSON, D. HARBATER & P. MÜLLER (eds)
257 An introduction to noncommutative differential geometry and its physical applications (2nd edition), J. MADORE
258 Sets and proofs, S.B. COOPER & J.K. TRUSS (eds)
259 Models and computability, S.B. COOPER & J. TRUSS (eds)
260 Groups St Andrews 1997 in Bath I, C.M. CAMPBELL *et al* (eds)
261 Groups St Andrews 1997 in Bath II, C.M. CAMPBELL *et al* (eds)
262 Analysis and logic, C.W. HENSON, J. IOVINO, A.S. KECHRIS & E. ODELL
263 Singularity theory, W. BRUCE & D. MOND (eds)
264 New trends in algebraic geometry, K. HULEK, F. CATANESE, C. PETERS & M. REID (eds)
265 Elliptic curves in cryptography, I. BLAKE, G. SEROUSSI & N. SMART
267 Surveys in combinatorics, 1999, J.D. LAMB & D.A. PREECE (eds)
268 Spectral asymptotics in the semi-classical limit, M. DIMASSI & J. SJÖSTRAND
269 Ergodic theory and topological dynamics of group actions on homogeneous spaces, M.B. BEKKA & M. MAYER
271 Singular perturbations of differential operators, S. ALBEVERIO & P. KURASOV
272 Character theory for the odd order theorem, T. PETERFALVI. Translated by R. SANDLING
273 Spectral theory and geometry, E.B. DAVIES & Y. SAFAROV (eds)
274 The Mandelbrot set, theme and variations, T. LEI (ed)
275 Descriptive set theory and dynamical systems, M. FOREMAN, A.S. KECHRIS, A. LOUVEAU & B. WEISS (eds)
276 Singularities of plane curves, E. CASAS-ALVERO
277 Computational and geometric aspects of modern algebra, M. ATKINSON *et al* (eds)
278 Global attractors in abstract parabolic problems, J.W. CHOLEWA & T. DLOTKO
279 Topics in symbolic dynamics and applications, F. BLANCHARD, A. MAASS & A. NOGUEIRA (eds)
280 Characters and automorphism groups of compact Riemann surfaces, T. BREUER
281 Explicit birational geometry of 3-folds, A. CORTI & M. REID (eds)
282 Auslander–Buchweitz approximations of equivariant modules, M. HASHIMOTO
283 Nonlinear elasticity, Y.B. FU & R.W. OGDEN (eds)
284 Foundations of computational mathematics, R. DEVORE, A. ISERLES & E. SÜLI (eds)
285 Rational points on curves over finite fields, H. NIEDERREITER & C. XING
286 Clifford algebras and spinors (2nd Edition), P. LOUNESTO
287 Topics on Riemann surfaces and Fuchsian groups, E. BUJALANCE, A.F. COSTA & E. MARTÍNEZ (eds)
288 Surveys in combinatorics, 2001, J.W.P. HIRSCHFELD (ed)
289 Aspects of Sobolev-type inequalities, L. SALOFF-COSTE
290 Quantum groups and Lie theory, A. PRESSLEY (ed)
291 Tits buildings and the model theory of groups, K. TENT (ed)
292 A quantum groups primer, S. MAJID
293 Second order partial differential equations in Hilbert spaces, G. DA PRATO & J. ZABCZYK
294 Introduction to operator space theory, G. PISIER
295 Geometry and integrability, L. MASON & Y. NUTKU (eds)
296 Lectures on invariant theory, I. DOLGACHEV
297 The homotopy category of simply connected 4-manifolds, H.-J. BAUES
298 Higher operads, higher categories, T. LEINSTER (ed)
299 Kleinian groups and hyperbolic 3-manifolds, Y. KOMORI, V. MARKOVIC & C. SERIES (eds)
300 Introduction to Möbius differential geometry, U. HERTRICH-JEROMIN
301 Stable modules and the D(2)-problem, F.E.A. JOHNSON
302 Discrete and continuous nonlinear Schrödinger systems, M.J. ABLOWITZ, B. PRINARI & A.D. TRUBATCH
303 Number theory and algebraic geometry, M. REID & A. SKOROBOGATOV (eds)
304 Groups St Andrews 2001 in Oxford I, C.M. CAMPBELL, E.F. ROBERTSON & G.C. SMITH (eds)
305 Groups St Andrews 2001 in Oxford II, C.M. CAMPBELL, E.F. ROBERTSON & G.C. SMITH (eds)
306 Geometric mechanics and symmetry, J. MONTALDI & T. RATIU (eds)
307 Surveys in combinatorics 2003, C.D. WENSLEY (ed.)
308 Topology, geometry and quantum field theory, U.L. TILLMANN (ed)

309 Corings and comodules, T. BRZEZINSKI & R. WISBAUER
310 Topics in dynamics and ergodic theory, S. BEZUGLYI & S. KOLYADA (eds)
311 Groups: topological, combinatorial and arithmetic aspects, T.W. MÜLLER (ed)
312 Foundations of computational mathematics, Minneapolis 2002, F. CUCKER *et al* (eds)
313 Transcendental aspects of algebraic cycles, S. MÜLLER-STACH & C. PETERS (eds)
314 Spectral generalizations of line graphs, D. CVETKOVIC, P. ROWLINSON & S. SIMIC
315 Structured ring spectra, A. BAKER & B. RICHTER (eds)
316 Linear logic in computer science, T. EHRHARD, P. RUET, J.-Y. GIRARD & P. SCOTT (eds)
317 Advances in elliptic curve cryptography, I.F. BLAKE, G. SEROUSSI & N.P. SMART (eds)
318 Perturbation of the boundary in boundary-value problems of partial differential equations, D. HENRY
319 Double affine Hecke algebras, I. CHEREDNIK
320 L-functions and Galois representations, D. BURNS, K. BUZZARD & J. NEKOVÁR (eds)
321 Surveys in modern mathematics, V. PRASOLOV & Y. ILYASHENKO (eds)
322 Recent perspectives in random matrix theory and number theory, F. MEZZADRI & N.C. SNAITH (eds)
323 Poisson geometry, deformation quantisation and group representations, S. GUTT *et al* (eds)
324 Singularities and computer algebra, C. LOSSEN & G. PFISTER (eds)
325 Lectures on the Ricci flow, P. TOPPING
326 Modular representations of finite groups of Lie type, J.E. HUMPHREYS
327 Surveys in combinatorics 2005, B.S. WEBB (ed)
328 Fundamentals of hyperbolic manifolds, R. CANARY, D. EPSTEIN & A. MARDEN (eds)
329 Spaces of Kleinian groups, Y. MINSKY, M. SAKUMA & C. SERIES (eds)
330 Noncommutative localization in algebra and topology, A. RANICKI (ed)
331 Foundations of computational mathematics, Santander 2005, L.M PARDO, A. PINKUS, E. SÜLI & M.J. TODD (eds)
332 Handbook of tilting theory, L. ANGELERI HÜGEL, D. HAPPEL & H. KRAUSE (eds)
333 Synthetic differential geometry (2nd Edition), A. KOCK
334 The Navier–Stokes equations, N. RILEY & P. DRAZIN
335 Lectures on the combinatorics of free probability, A. NICA & R. SPEICHER
336 Integral closure of ideals, rings, and modules, I. SWANSON & C. HUNEKE
337 Methods in Banach space theory, J.M.F. CASTILLO & W.B. JOHNSON (eds)
338 Surveys in geometry and number theory, N. YOUNG (ed)
339 Groups St Andrews 2005 I, C.M. CAMPBELL, M.R. QUICK, E.F. ROBERTSON & G.C. SMITH (eds)
340 Groups St Andrews 2005 II, C.M. CAMPBELL, M.R. QUICK, E.F. ROBERTSON & G.C. SMITH (eds)
341 Ranks of elliptic curves and random matrix theory, J.B. CONREY, D.W. FARMER, F. MEZZADRI & N.C. SNAITH (eds)
342 Elliptic cohomology, H.R. MILLER & D.C. RAVENEL (eds)
343 Algebraic cycles and motives I, J. NAGEL & C. PETERS (eds)
344 Algebraic cycles and motives II, J. NAGEL & C. PETERS (eds)
345 Algebraic and analytic geometry, A. NEEMAN
346 Surveys in combinatorics 2007, A. HILTON & J. TALBOT (eds)
347 Surveys in contemporary mathematics, N. YOUNG & Y. CHOI (eds)
348 Transcendental dynamics and complex analysis, P.J. RIPPON & G.M. STALLARD (eds)
349 Model theory with applications to algebra and analysis I, Z. CHATZIDAKIS, D. MACPHERSON, A. PILLAY & A. WILKIE (eds)
350 Model theory with applications to algebra and analysis II, Z. CHATZIDAKIS, D. MACPHERSON, A. PILLAY & A. WILKIE (eds)
351 Finite von Neumann algebras and masas, A.M. SINCLAIR & R.R. SMITH
352 Number theory and polynomials, J. MCKEE & C. SMYTH (eds)
353 Trends in stochastic analysis, J. BLATH, P. MÖRTERS & M. SCHEUTZOW (eds)
354 Groups and analysis, K. TENT (ed)
355 Non-equilibrium statistical mechanics and turbulence, J. CARDY, G. FALKOVICH & K. GAWEDZKI
356 Elliptic curves and big Galois representations, D. DELBOURGO
357 Algebraic theory of differential equations, M.A.H. MACCALLUM & A.V. MIKHAILOV (eds)
358 Geometric and cohomological methods in group theory, M.R. BRIDSON, P.H. KROPHOLLER & I.J. LEARY (eds)
359 Moduli spaces and vector bundles, L. BRAMBILA-PAZ, S.B. BRADLOW, O. GARCÍA-PRADA & S. RAMANAN (eds)
360 Zariski geometries, B. ZILBER
361 Words: Notes on verbal width in groups, D. SEGAL
362 Differential tensor algebras and their module categories, R. BAUTISTA, L. SALMERÓN & R. ZUAZUA
363 Foundations of computational mathematics, Hong Kong 2008, F. CUCKER, A. PINKUS & M.J. TODD (eds)
364 Partial differential equations and fluid mechanics, J.C. ROBINSON & J.L. RODRIGO (eds)
365 Surveys in combinatorics 2009, S. HUCZYNSKA, J.D. MITCHELL & C.M. RONEY-DOUGAL (eds)
366 Highly oscillatory problems, B. ENGQUIST, A. FOKAS, E. HAIRER & A. ISERLES (eds)
367 Random matrices: High dimensional phenomena, G. BLOWER
368 Geometry of Riemann surfaces, F.P. GARDINER, G. GONZÁLEZ-DIEZ & C. KOUROUNIOTIS (eds)
369 Epidemics and rumours in complex networks, M. DRAIEF & L. MASSOULIÉ
370 Theory of p-adic distributions, S. ALBEVERIO, A.YU. KHRENNIKOV & V.M. SHELKOVICH
371 Conformal fractals, F. PRZYTYCKI & M. URBAŃSKI
372 Moonshine: The first quarter century and beyond, J. LEPOWSKY, J. MCKAY & M.P. TUITE (eds)
373 Smoothness, regularity, and complete intersection, J. MAJADAS & A. RODICIO

London Mathematical Society Lecture Note Series: 374

Geometric Analysis of Hyperbolic Differential Equations: An Introduction

S. ALINHAC
Université Paris-Sud, Orsay

CAMBRIDGE
UNIVERSITY PRESS

CAMBRIDGE UNIVERSITY PRESS
Cambridge, New York, Melbourne, Madrid, Cape Town, Singapore,
São Paulo, Delhi, Dubai, Tokyo

Cambridge University Press
The Edinburgh Building, Cambridge CB2 8RU, UK

Published in the United States of America by Cambridge University Press, New York

www.cambridge.org
Information on this title: www.cambridge.org/9780521128223

First published 2010

Printed in the United Kingdom at the University Press, Cambridge

A catalogue record for this publication is available from the British Library

Library of Congress Cataloguing in Publication data
Alinhac, S. (Serge)
Geometric analysis of hyperbolic differential equations : an introduction / S. Alinhac.
p. cm. – (London Mathematical Society lecture note series ; 374)
Includes bibliographical references and index.
ISBN 978-0-521-12822-3 (pbk.)
1. Nonlinear wave equations. 2. Differential equations, Hyperbolic. 3. Quantum theory.
4. Geometry, Differential. I. Title. II. Series.
QA927.A3886 2010
515′.3535 – dc22 2010001099

ISBN 978-0-521-12822-3 Paperback

Contents

	Preface	*page* vii
1	**Introduction**	1
2	**Metrics and frames**	8
2.1	Metrics, duality	8
2.2	Optical functions	12
2.3	Null frames	12
3	**Computing with frames**	17
3.1	Metric connexion	17
3.2	Submanifolds	20
3.3	Hessian and d'Alembertian	21
3.4	Frame coefficients	24
4	**Energy inequalities and frames**	29
4.1	The energy–momentum tensor	29
4.2	Deformation tensor	31
4.3	Energy inequality formalism	33
4.4	Energy	34
4.5	Interior terms and positive fields	35
4.6	Maxwell equations	41
	4.6.1 Duality	41
	4.6.2 Energy formalism	43
5	**The good components**	45
5.1	The problem	45
5.2	An important remark	46
5.3	Ghost weights and improved standard energy inequalities	47
5.4	Conformal inequalities	53

6	**Pointwise estimates and commutations**	57
6.1	Pointwise decay and conformal inequalities	58
6.2	Commuting fields in the scalar case	59
6.3	Modified Lorentz fields	61
6.4	Commuting fields for Maxwell equations	63
7	**Frames and curvature**	65
7.1	The curvature tensor	65
7.2	Optical functions and curvature	67
7.3	Transport equations	69
7.4	Elliptic systems	71
7.5	Mixed transport–elliptic systems	75
8	**Nonlinear equations, a priori estimates and induction**	77
8.1	A simple ODE example	77
8.2	Local existence theory	80
8.3	Blowup criteria	81
8.4	Induction on time for PDEs	84
9	**Applications to some quasilinear hyperbolic problems**	88
9.1	Quasilinear wave equations satisfying the null condition	89
9.2	Quasilinear wave equations	96
9.3	Low regularity well-posedness for quasilinear wave equations	99
9.4	Stability of Minkowski spacetime (first version)	102
9.5	L^2 conjecture on the curvature	106
9.6	Stability of Minkowski spacetime (second version)	108
9.7	The formation of black holes	113
	References	114
	Index	117

Preface

The field of nonlinear hyperbolic equations or systems has seen a tremendous development since the beginning of the 1980s. We are concentrating here on multidimensional situations, and on quasilinear equations or systems, that is, when the coefficients of the principal part depend on the unknown function itself. The pioneering works by F. John, D. Christodoulou, L. Hörmander, S. Klainerman, A. Majda and many others have been devoted mainly to the questions of blowup, lifespan, shocks, global existence, etc. Some overview of the classical results can be found in the books of Majda [42] and Hörmander [24]. On the other hand, Christodoulou and Klainerman [18] proved in around 1990 the stability of Minkowski space, a striking mathematical result about the Cauchy problem for the Einstein equations. After that, many works have dealt with diagonal systems of quasilinear wave equations, since this is what Einstein equations reduce to when written in the so-called harmonic coordinates. The main feature of this particular case is that the (scalar) principal part of the system is a wave operator associated to a unique Lorentzian metric on the underlying space-time. This is in strong contrast with the more complicated case of general symmetric quasilinear systems: the compressible isentropic Euler equations, for instance, can be viewed as a quasilinear wave equation coupled to a vector field; the system of nonlinear elasticity involves two different wave equations, etc.

I consider here only the case of quasilinear wave equations. We observe two main domains of interest: the study of *global* smooth solutions, and the study of *low regularity* solutions, both domains being connected. The striking feature is the *unity* of the techniques and ideas used in the works on these domains: The emphasis is always on good directions and good components, these components being components of tensors relative to some special frames, the *null frames*. Hence the observed unity comes from the fact that most concepts, such as metrics, connexions, curvature, etc., are borrowed from Lorentzian geometry. This is, of course, related to mathematical work by Penrose and collaborators in the domain of general relativity, where null frames have been used extensively (see, for instance, Penrose and Rindler [43]).

Since the work of Christodoulou and Klainerman cited above, many mathematical papers on the subject of quasilinear wave equations or Einstein equations use the language of Lorentzian geometry and deal with energy–momentum tensors, deformation tensors, etc. However, there seem to be some difficulties: Riemannian geometry books do not include the specific Lorentzian tools such as null frames; most relativity books do not include a description of the relevant mathematical techniques. Let us, however, draw attention to the books of Hawking and Ellis [23] and Rendall [44]; which include substantial mathematics.

I believe that the use of Lorentzian tools in the mathematical study of nonlinear hyperbolic systems is going to intensify further, even in the aspects of the field not directly related with general relativity. This is what we call "geometric analysis of hyperbolic equations." It is true that there are examples of nonlinear wave equations which are perturbations of the standard wave equation by small nonlinear terms, where it is enough to consider only the geometry of the standard wave equation, that is, the Minkowski metric: These examples are striking, but the possibility of this simplification seems to be related to the fact that one is considering only *small* solutions; for large solutions, we believe that it will be necessary to take into account the geometry of the linearized operator, that is, a Lorentzian metric depending on the solution itself.

This book is meant for people wanting to access the mathematical literature on the subject of quasilinear wave equations or Einstein equations. Its goal is twofold:

(i) To give to analysts in the field of partial differential equations (PDEs) a self-contained and elementary access to the necessary tools of Lorentzian geometry,
(ii) To explain the fundamental ideas connected with the use of null frames.

This book can be read by students or researchers with an elementary background in distribution theory and linear PDEs, specifically hyperbolic PDEs. No knowledge of differential geometry is required. Though the largest part of the text is about geometric concepts, this book is not a book about Lorentzian geometry: it introduces the geometric tools required to understand the modern PDE literature only as and when they are needed. The author not being a geometer, I deliberately chose to give naive and self-contained proofs to all statements, which can be viewed as "do it yourself" exercises for the reader, without using sophisticated "well-known" facts. I hope that I will be forgiven for that.

Finally, I would like to thank S. Klainerman and F. Labourie for many helpful conversations.

1
Introduction

The prototype of all hyperbolic equations is the wave equation, or d'Alembertian

$$\Box \equiv \partial_t^2 - \Delta_x, \ \Delta = \partial_1^2 + \partial_2^2 + \partial_3^2$$

in $\mathbf{R}^4_{x,t}$. In order to introduce the concepts and questions of this book, we first review briefly some decay properties of the solutions of $\Box\phi = 0$. We refer the reader to [9] for all formula and proofs.

1. We consider in $\mathbf{R}^4_{x,t}$ the Cauchy problem for the standard wave equation

$$\Box\phi = (\partial_t^2 - \Delta_x)\phi = 0, \ \phi(x,0) = \phi_0(x), \ (\partial_t\phi)(x,0) = \phi_1(x).$$

a. Suppose for simplicity $\phi_0, \phi_1 \in C_0^\infty$, $\phi_i(x) = 0$ for $r = |x| \geq M$: as a consequence of the classical solution formula, the function ϕ can be represented for $r \geq 1$ as

$$\phi(x,t) = \frac{1}{r} F\left(r - t, \omega, \frac{1}{r}\right), \ r = |x|, \ \omega = \frac{x}{r}, \ \sigma = r - t, \ z = \frac{1}{r},$$

for some C^∞ function $F(\sigma, \omega, z)$. Since the propagation speed is 1, F vanishes for $r \geq t + M$, that is, $\sigma = r - t \geq M$. By the strong Huygens principle, the solution also vanishes for $r \leq t - M$, that is, $\sigma = r - t \leq -M$. Thus F and ϕ are supported in the strip $|r - t| \leq M$ close to the light cone $\{r = t\}$. Setting $\partial_r = \sum \omega^i \partial_i$, we introduce the two fields

$$L = \partial_t + \partial_r, \ \underline{L} = \partial_t - \partial_r,$$

and define the rotation fields $R = x \wedge \partial$,

$$R_1 = x^2\partial_3 - x^3\partial_2, \ R_2 = x^3\partial_1 - x^1\partial_3, \ R_3 = x^1\partial_2 - x^2\partial_1.$$

Note that $R_i(r) = 0$, and $\sum \omega^i R_i = 0$. Using the representation formula, we observe that

$$L\phi = \frac{-F}{r^2} - \frac{\partial_z F}{r^3} = O(r^{-2}),\ r \to +\infty.$$

Similarly, since $\partial_i \omega^j = (\delta_i^j - \omega^i \omega^j)/r$,

$$\frac{R}{r}\phi = O(r^{-2}),\ r \to +\infty,$$

while, for instance, $\underline{L}\phi$ has only magnitude r^{-1}. Hence the special derivatives $L\phi$, $(R/r)\phi$ behave better at infinity than the other components of $\nabla\phi$. We call them the "good derivatives" of ϕ.

b. We explain now how to obtain the same decay result for the good derivatives using an "energy method," which is an alternative approach to the preceding decay results that does not use an explicit representation for ϕ. We define the hyperbolic rotations $H = t\partial + x\partial_t$,

$$H_1 = t\partial_1 + x^1\partial_t,\ H_2 = t\partial_2 + x^2\partial_t,\ H_3 = t\partial_3 + x^3\partial_t,$$

and call Lorentz fields Z all the fields

$$\partial_\alpha,\ S = t\partial_t + \sum x^i \partial_i = t\partial_t + r\partial_r,\ R = x \wedge \partial,\ H = t\partial + x\partial_t.$$

These fields are known to commute with $\Box$, except for the scaling field S which satisfies $[\Box, S] = 2\Box$. In the situation in **a**, commuting the fields Z with $\Box$ we obtain $\Box Z\phi = 0$; using the standard energy inequality for the wave equation, we obtain the bound

$$\sum ||(\nabla Z\phi)(\cdot, t)||_{L^2_x} \leq C.$$

Now, the following easy formula establishes a connexion between the special derivatives L, R/r and the Lorentz fields:

$$(r+t)L = S + \sum \omega_i H_i,\ (t-r)\underline{L} = S - \sum \omega_i H_i,\ \frac{R}{r} = t^{-1}\omega \wedge H.$$

Note also that for any smooth function supported in $|r - t| \leq M$, we have the Poincaré inequality

$$||w(\cdot, t)||_{L^2} \leq C||(\partial_r w)(\cdot, t)||_{L^2}.$$

Using these formulas, we get for the special derivatives of $L\phi$, $(R/r)\phi$

$$||(\nabla L\phi)(\cdot, t)||_{L^2} = O(t^{-1}),\ \left\| \nabla \frac{R}{r}\phi(\cdot, t) \right\|_{L^2} = O(t^{-1}),\ t \to +\infty.$$

Taking into account the support of ϕ and using again the Poincaré inequality, we even obtain

$$||(L\phi)(\cdot, t)||_{L^2} = O(t^{-1}), \ \left\|\frac{R}{r}\phi(\cdot, t)\right\|_{L^2} = O(t^{-1}), \ t \to +\infty.$$

Note the contrast with the information given by the standard energy inequality, which yields only the boundedness of these quantities.

It is, in fact, possible to recover the pointwise estimates from **a** using the preceding L^2-estimates. For this, we first commute a product Z^k of k of the Lorentz fields with the wave equation, thus obtaining $\Box Z^k\phi = 0$. Then we use the Klainerman inequality, which is valid for any smooth function v sufficiently decaying at infinity:

$$|v(x, t)|(1 + t + r)(1 + |t - r|)^{\frac{1}{2}} \leq C \sum_{k \leq 2} ||Z^k v(\cdot, t)||_{L^2}.$$

We thus obtain again the pointwise bounds that we had from the explicit representation formula

$$L\phi = O(t^{-2}), \ \frac{R}{r}\phi = O(t^{-2}).$$

Note, however, that this "energy method" is likely to work in variable coefficients situations (or nonlinear situations), where we do not know the representation formula.

If the data are not compactly supported but are sufficiently decaying as $|x| \to +\infty$, this energy method still works, but the "interior" behavior of the solution (that is, away from the light cone $\{t = r\}$) is not as good as before.

c. In **b**, we commuted products of Lorentz fields with $\Box$ and then used the *standard* energy inequality. There is, however, still another type of "energy approach" that displays better behavior of the special derivatives $L\phi$, $(R/r)\phi$. This approach does not involve Lorentz fields, but instead requires a different type of energy inequality. We give two examples of this.

First, one can prove the following improvement of the standard energy inequality: for all $\epsilon > 0$, there is some constant $C_\epsilon > 0$ such that, assuming $\Box\phi = 0$,

$$E_\phi(T)^{\frac{1}{2}} + \left\{\int_{0 \leq t \leq T} \langle r - t\rangle^{-1-\epsilon} \left[(L\phi)^2 + \left|\frac{R}{r}\phi\right|^2\right] dxdt\right\}^{\frac{1}{2}} \leq C_\epsilon E_\phi(0)^{\frac{1}{2}}.$$

Here, E_ϕ is the standard energy

$$E_\phi(t) = \tfrac{1}{2}\int [(\partial_t\phi)^2 + |\nabla_x\phi|^2](x, t)dx.$$

This inequality is easily obtained in the same way as the usual energy inequality, using the multiplier ∂_t and a weight e^a, where $a = a(r-t)$ is appropriately chosen (see [9], for instance). This inequality is only useful in a region where $|r-t|$ is smaller than t, that is, close to the light cone. In the region $|r-t| \leq C$ for instance, the L^2_x norm of the special derivatives $L\phi$, $(R/r)\phi$ is not just bounded, it is an L^2 function of t. We can thus identify the "good derivatives" of ϕ directly from the energy inequality, without commuting any fields with the equation.

The second example of an inequality displaying the good derivatives is the conformal energy inequality which gives, for $\Box\phi = 0$,

$$\tilde{E}_\phi(t)^{\frac{1}{2}} \leq C\tilde{E}_\phi(0)^{\frac{1}{2}},$$

where the conformal energy $\tilde{E}$ is

$$\tilde{E}_\phi(t) = \tfrac{1}{2}\int [(S\phi)^2 + |R\phi|^2 + |H\phi|^2 + \phi^2](x,t)dx.$$

This inequality is obtained in the usual way using the timelike multiplier K_0:

$$K_0 = (r^2+t^2)\partial_t + 2rt\partial_r.$$

Using the identities $(r+t)L = S + \sum \omega_i H_i$, $R/r = t^{-1}\omega \wedge H$ from **b**, the bound of the quantities $||(Z\phi)(\cdot,t)||_{L^2}$ provided by the inequality yields the bounds

$$||(L\phi)(\cdot,t)||_{L^2} = O(t^{-1}), \quad \left\| \frac{R}{r}\phi(\cdot,t)\right\|_{L^2} = O(t^{-1}).$$

Once again, we can identify the good derivatives of ϕ directly from the conformal energy inequality.

2. Consider now, at each point away from $r = 0$, the null frame

$$e_1,\ e_2,\ e_3 = \underline{L} = \partial_t - \partial_r,\ e_4 = L = \partial_t + \partial_r,$$

where, at each point (x_0, t_0), (e_1, e_2) form an orthonormal basis of the tangent space to the sphere

$$\{(x,t),\ t = t_0,\ |x| = |x_0|\}.$$

Using spherical coordinates

$$x^1 = r\sin\theta\cos\phi,\ x^2 = r\sin\theta\sin\phi,\ x^3 = r\cos\theta,$$

we can take (away from the poles)

$$e_1 = r^{-1}\partial_\theta,\ e_2 = (r\sin\theta)^{-1}\partial_\phi.$$

The fields e_1, e_2 are related to the rotation fields by the formulas

$$e_1 = -(\sin\phi)\frac{R_1}{r} + (\cos\phi)\frac{R_2}{r}, \; e_2 = (\sin\theta)^{-1}\frac{R_3}{r}.$$

Hence the "special derivatives" of ϕ on which we insisted above are just, equivalently, the components of $d\phi$ on e_1, e_2, and L, that is, some of the components of $d\phi$ in a null frame, the only bad derivative being $\underline{L}\phi$.

To understand the name "null frame," it is best to introduce on $\mathbf{R}^4$ the scalar product of special relativity. For two vectors $X = (X^0, X^1, X^2, X^3)$ and $Y = (Y^0, Y^1, Y^2, Y^3)$, we set

$$\langle X, Y\rangle = -X^0Y^0 + \sum_{1\le i\le 3} X^iY^i.$$

We can then easily check the fundamental properties which define a null frame:

$$(e_1, e_2) \perp (e_3, e_4), \; \langle L, L\rangle = 0, \; \langle \underline{L}, \underline{L}\rangle = 0, \; \langle L, \underline{L}\rangle = -2.$$

The "gradient" $\tilde{\nabla} f$ of a function f in the sense of this scalar product is defined by

$$\forall Y, \; \langle \tilde{\nabla} f, Y\rangle = df(Y) = Y(f).$$

This gives immediately

$$\tilde{\nabla} f = (-\partial_t f, \partial_1 f, \partial_2 f, \partial_3 f).$$

For instance,

$$\tilde{\nabla}(t-r) = -\left(1, \frac{x}{r}\right) = -L.$$

Since L is "null," we also have, with $u = t - r$,

$$\langle \tilde{\nabla} u, \tilde{\nabla} u\rangle = 0,$$

and we say that u is an optical function. Note that the null frame $(e_1, e_2, \underline{L}, L)$ is associated to the functions u and t in the sense that:

(i) the surfaces $\{t = t_0, \; u = u_0\}$ are the usual spheres,
(ii) $L = -\tilde{\nabla} u$ and $(\underline{L}, L)$ are the two null vectors in the orthogonal space to these spheres.

This shows us how null frames and optical functions are related. Of course, the function $\underline{u} = t + r$ is also an optical function, and $\underline{L} = -\tilde{\nabla}\underline{u}$. Note that the level surfaces of u are outgoing light cones, while level surfaces of $\underline{u}$ are incoming light cones; also, the good derivatives (e_1, e_2, L) span, at each point, the tangent space to the outgoing cone through this point.

Let us mention to finish the relations between the fields S, K_0 (that we have already encountered) and u, L, $\underline{u}$, $\underline{L}$,

$$S = \frac{1}{2}(u\underline{L} + \underline{u}L),\ K_0 = \frac{1}{2}(u^2\underline{L} + \underline{u}^2 L).$$

3. The aim of this book is to explain how one can extend the previously discussed concepts and results to a general framework. More precisely, suppose we have, instead of the "flat" Minkowski metric $|X|^2 = \langle X, X\rangle$, a more general metric g:

$$g = \sum g_{\alpha\beta} dx^\alpha dx^\beta,\ g(X, Y) \equiv \langle X, Y\rangle = \sum g_{\alpha\beta} X^\alpha Y^\beta.$$

We assume, of course, that this metric has the signature $-, +, +, +$ just like the Minkowski metric. We define the wave equation $\square$ associated with this metric by

$$\square_g \phi \equiv \square\phi = |g|^{-\frac{1}{2}} \sum \partial_\alpha (g^{\alpha\beta} |g|^{\frac{1}{2}} \partial_\beta \phi),$$

where $|g|$ is the determinant of the matrix $(g_{\alpha\beta})$ and $(g^{\alpha\beta})$ its inverse matrix. However, we sometimes write $\square = \partial_t^2 - \Delta$ for the standard wave operator (instead of $-\partial_t^2 + \Delta$). Our interest centres on these wave equations, and also on the associated Maxwell and Bianchi equations. From the considerations above for the "flat" case of the Minkowski metric, the following natural questions arise: for solutions ϕ of $\square_g \phi = 0$,

(i) Are there "good derivatives" of ϕ (in the sense of a better decay at infinity), analogous to $L\phi$, $(R/r)\phi$?
(ii) How should a null frame that captures these "good derivatives" be chosen?
(iii) What is the relation between null frames and optical functions?
(iv) Can one prove energy inequalities where the good derivatives are singled out, as in **1.c**?
(v) Are there good substitute for the Lorentz fields Z?
(vi) Can one commute these substitutes with $\square$ to obtain pointwise bounds for the solutions, as in **1.b**?

The plan of the book follows from what we said before, about introducing the necessary geometric machinery only as and when it is needed.

- In chapter 2, we discuss the notions of metric, optical functions, and null frames, and give simple examples found in the literature.
- The differential geometry aspects appear in chapter 3, where the metric connexion is introduced, as a necessary tool to deal with frames; we then define the frame coefficients and compute them for simple examples.

- Chapter 4 is dedicated to the specific machinery used to prove energy inequalities: the energy–momentum tensor, the deformation tensor, etc. The idea is to do the computations in such a way that the energy and the additional "interior terms" can be easily expressed in the frame in which we are working.
- The question of how to chose a good frame and thus identify the good components of tensors is adressed in chapter 5, where we discuss extensions of the standard energy inequality and of the conformal energy inequality.
- The way to find substitutes for the standard Lorentz fields and to commute them with $\Box$ is explained in chapter 6.
- The curvature tensor is introduced only in chapter 7, where we explain how to control optical functions and their associated null frames. We establish there the transport equations and elliptic systems (on (nonstandard) 2-spheres) which govern the frame coefficients.
- Finally, the last two chapters are devoted to discussing a number of applications of the ideas presented in the previous chapters to nonlinear problems. Though it seems impossible to give complete proofs of very difficult results, we try to outline the constructions of frames, the inequalities used, etc., in the hope of providing a guide for further reading.

2

Metrics and frames

2.1 Metrics, duality

1. We will work in $\mathbf{R}^4$ or in a four-dimensional manifold M. Local coordinates on M are denoted by x^α, $\alpha = 0, 1, 2, 3$. Sometimes, $x^0 = t$ is thought of as "the time," while (x^1, x^2, x^3) are the "spatial coordinates," though this does not make much sense in the context of relativity theory. The corresponding partial derivatives are $\partial_\alpha = \partial/\partial x^\alpha$. From now on, we assume the concepts of vector fields and 1-forms are known, referring to [46] if necessary.

The **position of the indices** is crucial: vector fields are indexed with a lower index, such as ∂_α, 1-forms are indexed with an upper index, such as dx^α. The components of a vector field X are denoted by X^α, since $X^\alpha = dx^\alpha(X)$, and the components of a 1-form ω are denoted by ω_α, since $\omega_\alpha = \omega(\partial_\alpha)$. Here and in what follows, a *repeated* sum on an index in the lower and the upper position is *never* indicated; for instance, we write in local coordinates a vector field $X = \sum X^\alpha \partial_\alpha = X^\alpha \partial_\alpha$, a 1-form $\omega = \sum \omega_\alpha dx^\alpha = \omega_\alpha dx^\alpha$. If f is a function, we define $df = \sum (\partial_\alpha f) dx^\alpha = (\partial_\alpha f) dx^\alpha$, and $Xf = X^\alpha \partial_\alpha f$, etc.

2. A **metric** is the smooth assignment to each point m of a symmetric non-degenerate bilinear form on $T_m M$. In local coordinates x^α, the components of the metric are $g_{\alpha\beta} = g(\partial_\alpha, \partial_\beta)$, which are supposed to be smooth. Hence g can be locally identified with the symmetric 4×4 invertible matrix $(g_{\alpha\beta})$. The elements of the inverse matrix are denoted by $g^{\alpha\beta}$, the determinant of g by $|g|$. Throughout the book, the signature of the quadratic form g will be $(-1, +1, +1, +1)$; in other words, the metric is assumed to be Lorentzian, in contrast with the Riemanian case, where g is assumed to be positive definite.

Using the same convention on repeated indices, the metric is sometimes written

$$g \equiv ds^2 = g_{\alpha\beta} dx^\alpha dx^\beta, \; g(X, Y) = \langle X, Y \rangle = g_{\alpha\beta} X^\alpha Y^\beta.$$

We give a short list of the most common **examples of Lorentzian metrics**:

(i) The **Minkowski metric** (also called "flat" metric) is given on $\mathbf{R}^4_{x,t}$ by

$$g = -dt^2 + (dx^1)^2 + (dx^2)^2 + (dx^3)^2.$$

Using spherical coordinates (see the introduction), we get

$$g = -dt^2 + dr^2 + r^2(d\theta^2 + (\sin^2\theta)d\phi^2).$$

Setting $\underline{u} = t + r$, $u = t - r$, then $\tan p = \underline{u}$, $\tan q = u$, it is often convenient to compactify the whole of $\mathbf{R}^4$ by introducing new coordinates $t' = p + q$, $r' = p - q$, with

$$-\pi < t' + r' < \pi, \ -\pi < t' - r' < \pi, \ r' \geq 0.$$

In these coordinates, the metric is

$$g = [4\cos^2(\tfrac{1}{2}(t' + r'))\cos^2(\tfrac{1}{2}(t' - r'))]^{-1}\tilde{g},$$

$$\tilde{g} = -dt'^2 + dr'^2 + \sin^2 r'(d\theta^2 + (\sin^2\theta)d\phi^2).$$

The corresponding drawing in two dimensions in the coordinates (r', t') is called a Penrose diagram. It allows a better understanding of "infinity": the lines $\mathcal{I}^+ = \{r' + t' = \pi\}$ and $\mathcal{I}^- = \{t' - r' = -\pi\}$ are called, respectively, future and past null infinity, the point $(r' = \pi, t' = 0)$ spatial infinity, etc.

(ii) Perturbations of the Minkowski metric can be defined by

$$g = -dt^2 + \sum g_{ij}dx^i dx^j.$$

Note that in this context, latin indices run from 1 to 3 (while greek indices run from 0 to 3). The matrix g_{ij} is assumed to be positive definite. As there are no "cross terms", we say that g is **split**.

(iii) The **Schwarzschild metric** is

$$g = -\left(1 - \frac{2m}{r}\right)dt^2 + \left(1 - \frac{2m}{r}\right)^{-1} dr^2 + r^2(d\theta^2 + (\sin^2\theta)d\phi^2).$$

Here, $m \geq 0$ is given, and (r, θ, ϕ) are spherical coordinates on $\mathbf{R}^3$. When $m = 0$, this metric reduces to the Minkowski metric written in spherical coordinates. One can show that the surface $\{r = 2m\}$ is only an *apparent* singularity of the metric, and one can also construct Penrose diagrams for this metric (see [23] for details).

(iv) The **Kerr metric** is

$$g = -\frac{\Delta - a^2 \sin^2 \theta}{\Sigma} dt^2 + \frac{\Sigma}{\Delta} dr^2 - 4amr \frac{\sin^2 \theta}{\Sigma} dt d\phi$$
$$+ A \sin^2 \theta d\phi^2 + \Sigma d\theta^2,$$

with $\Sigma = r^2 + a^2 \cos^2 \theta$, $\Delta = r^2 + a^2 - 2mr$, $A = (r^2 + a^2)^2 - a^2 \Delta \sin^2 \theta$. When $a = 0$, g is the Schwarzschild metric. Again, how to construct Penrose diagrams for Kerr metrics is explained in [23].

3. A given metric provides a bijection between vector fields and 1-forms according to the formula

$$\forall Y, \ \langle X, Y \rangle = \omega(Y).$$

We say that X and ω are **dual** to each other if the above relation is true for all vector fields Y. In local coordinates, this reads

$$X = X^\alpha \partial_\alpha, \ \omega = \omega_\alpha dx^\alpha, \ g_{\alpha\beta} X^\beta = \omega_\alpha.$$

We say that ω_α is obtained from X^β by "lowering" the index, and we just write $X_\alpha = g_{\alpha\beta} X^\beta$. Analogously, we write, "raising" the index, $\omega^\alpha = g^{\alpha\beta} \omega_\beta$. Hence we do not distinguish between X and ω, using the same letter for both.

Recall that for each point m, a p-tensor T is a p-multilinear form on $T_m M$ (depending smoothly on m). The case $p = 1$ corresponds to 1-forms. For a given basis (e_α), the components of T in this basis are

$$T_{\alpha\beta\ldots\gamma} = T(e_\alpha, e_\beta, \ldots, e_\gamma).$$

If we have chosen local coordinates x^α, it is understood that we take $e_\alpha = \partial_\alpha$, thus

$$T_{\alpha\beta\ldots\gamma} = T(\partial_\alpha, \partial_\beta, \ldots, \partial_\gamma).$$

For any tensor T, the process of raising or lowering indices is the same as before. For instance, for a 2-tensor T, we set

$$T_\alpha^\beta = g^{\beta\gamma} T_{\gamma\alpha},$$

and so on.

a. Let f be a C^1 function on M. The **gradient** ∇f is defined as the dual of df, with components

$$\nabla f^\alpha = \partial^\alpha f \equiv g^{\alpha\beta} \partial_\beta f.$$

Note that, by definition, $\langle \nabla f, X \rangle = df(X) = Xf$, a very useful formula.

b. If (e_α) is a basis, we denote by $g_{\alpha\beta}$ the components of the symmetric 2-tensor g in this basis, and $g^{\alpha\beta}$ is as before the inverse matrix to $g_{\alpha\beta}$. The **dual basis** of (e_α) is then defined to be

$$e^\alpha = g^{\alpha\beta} e_\beta,$$

in such a way that

$$\langle e^\alpha, e_\beta \rangle = g^{\alpha\gamma} \langle e_\gamma, e_\beta \rangle = g^{\alpha\gamma} g_{\gamma\beta} = \delta^\alpha_\beta,$$

δ^β_α being 1 for $\alpha = \beta$ and 0 otherwise. For a 2-tensor $T_{\alpha\beta}$, we define its **trace** to be

$$\mathrm{tr} T = g^{\alpha\beta} T_{\alpha\beta} = T(e_\alpha, e^\alpha) = T(e^\alpha, e_\alpha).$$

The remarkable fact is that this trace is independent of the basis e_α chosen. In fact, if we denote again abusively the matrix with elements $T_{\alpha\beta}$ by T, the definition gives

$$\mathrm{tr} T = \mathrm{trace}\,(T g^{-1}).$$

Let us now consider a new basis

$$\tilde{e}_\alpha = A^\beta_\alpha e_\beta.$$

The new matrix $\tilde{T}$ is now

$$\tilde{T}_{\alpha\beta} = T(\tilde{e}_\alpha, \tilde{e}_\beta) = A^{\alpha'}_\alpha A^{\beta'}_\beta T_{\alpha'\beta'}.$$

Denoting the matrix $a_{\alpha\beta} = A^\beta_\alpha$ by a, we see that $\tilde{T} = aT^t a$. Similarly, $\tilde{g} = ag^t a$. Hence

$$\mathrm{trace}\,(\tilde{T}\tilde{g}^{-1}) = \mathrm{trace}\,(aTg^{-1}a^{-1}) = \mathrm{trace}\,(Tg^{-1}),$$

which gives the desired invariance.

c. Define locally the 4-form ϵ to be

$$\epsilon = |g|^{\frac{1}{2}} dx^0 \wedge dx^1 \wedge dx^2 \wedge dx^3.$$

From now on, we assume that we are working on $\mathbf{R}^4$ or on an *orientable* manifold M. One can then easily check that ϵ does not depend on the local coordinates; we call this the **volume form**. One should not confuse this volume form with the volume element used to integrate functions on the manifold, which is the positive measure $dv = |g|^{\frac{1}{2}} dx$ (see [46]).

d. For any tensor $T_{\alpha\beta\ldots\gamma}$, we define

$$|T|^2 = T_{\alpha\beta\cdots\gamma} T^{\alpha\beta\cdots\gamma}.$$

For instance, if T is a vector field, or a 1-form,

$$|T|^2 = T_\alpha T^\alpha = g_{\alpha\beta} T^\beta T^\alpha = \langle T, T \rangle$$

as expected. Note that $|T|^2$ has no reason to be nonnegative.

2.2 Optical functions

A C^1 function u is called an **optical function** if it satisfies the **eikonal equation**

$$g^{\alpha\beta} \partial_\alpha u \partial_\beta u = g_{\alpha\beta} \partial^\alpha u \partial^\beta u = \langle \nabla u, \nabla u \rangle = |\nabla u|^2 = 0.$$

In PDE terms, this means that the level surfaces $\{u = C\}$ are characteristic surfaces for any operator with principal symbol $g^{\alpha\beta}\xi_\alpha\xi_\beta$. The classical examples for the Minkowski metric are the functions $u = t - r$ and $\underline{u} = t + r$; in this case, the level surfaces $\{u = C\}$ and $\{\underline{u} = C\}$ are respectively, the outgoing light cones and the incoming light cones with vertices on the t-axis, and the integral curves of the fields $L = -\nabla u$ and $\underline{L} = -\nabla \underline{u}$ generate the cones. In general, note that if an optical function u is constant on a surface S, then ∇u is both normal and tangent to S. Conversely, if u is constant on a surface S to which ∇u is tangent, this implies that u is an optical function.

There are many ways of constructing optical functions for a given metric. One possibility is to solve a Cauchy problem for the eikonal equation $g^{\alpha\beta}\partial_\alpha u \partial_\beta u = 0$ with data on some hypersurface, using the classical method of characteristics. Another possibility is to define first outgoing and incoming (half-)cones with vortices on some line, and to take u and $\underline{u}$ to be functions having these half-cones as level surfaces. We will come back to this in section 3.3 and in the last chapter. Below, we show the role played by optical functions in constructing the special basis called null frames.

2.3 Null frames

In a Euclidean space, the orthonormal basis play an important role. The corresponding concept for a Lorentzian metric is that of **null frame**. A null frame is a basis (e_1, e_2, e_3, e_4) given at each point (and depending smoothly on this point), such that

$$\langle e_1, e_1 \rangle = 1,\ \langle e_2, e_2 \rangle = 1,\ \langle e_1, e_2 \rangle = 0,$$
$$\langle e_3, e_3 \rangle = 0,\ \langle e_4, e_4 \rangle = 0,\ \langle e_3, e_4 \rangle = -2\mu,$$

and the subspace generated by (e_1, e_2) is orthogonal to the subspace generated by (e_3, e_4). The vectors e_3 and e_4 are called null vectors (that is, they are isotropic for the quadratic form g). We always arrange $\mu > 0$, and sometimes, in fact, $\mu \equiv 1$.

The most classical example of null frame is, for the Minkowski metric, using spherical coordinates,

$$e_1 = r^{-1}\partial_\theta,\ e_2 = (\sin\theta)^{-1}\partial_\phi,\ e_3 = \underline{L} = \partial_t - \partial_r,\ e_4 = L = \partial_t + \partial_r.$$

The first two vectors form an orthonormal basis on the spheres of $\mathbf{R}^3_x$ (for constant t), the last two a basis of the orthogonal space to the sphere, and $\mu = 1$.

For the Schwarzschild metric, since the metric is rotationally invariant, we also use the standard spheres, the restriction of the metric being then the standard Euclidean metric. We take (e_1, e_2) as usual, and

$$e_3 = \partial_t - \left(1 - \frac{2m}{r}\right)\partial_r,\ e_4 = \partial_t + \left(1 - \frac{2m}{r}\right)\partial_r.$$

The case of the Kerr metric is more delicate, since the metric is not rotationally invariant. According to [47], there are good *algebraic* reasons to set

$$X = \partial_t + \frac{a}{r^2 + a^2}\partial_\phi,\ Y = \frac{\Delta}{r^2 + a^2}\partial_r,$$

and to choose

$$e_1 = \Sigma^{-\frac{1}{2}}\partial_\theta,\ e_2 = (\Sigma^{\frac{1}{2}}\sin\theta)^{-1}(\partial_\phi + a(\sin^2\theta)\partial_t),$$

$$e_3 = X - Y,\ e_4 = X + Y,\ \langle e_3, e_4\rangle = -2\Sigma\Delta(r^2 + a^2)^{-2}.$$

Note that (e_1, e_2) are not tangent to a sphere foliation, since

$$[e_1, e_2] = (r^2 + a^2)\Sigma^{-\frac{3}{2}}\cos\theta[a\Sigma^{-\frac{1}{2}}(e_3 + e_4) - (\sin\theta)^{-1}e_2]$$

is not generated by (e_1, e_2).

The matrix $g_{\alpha\beta}$ of the components of g in a null frame is split into two 2-blocks, and, for $X = X^\alpha e_\alpha$,

$$g(X, X) = (X^1)^2 + (X^2)^2 - 4\mu X^3 X^4.$$

Note, also, for further reference the formula which gives ∇f in a null frame:

$$\nabla f = e_1(f)e_1 + e_2(f)e_2 - \frac{1}{2\mu}(e_4(f)e_3 + e_3(f)e_4).$$

The dual basis of a null frame is the basis

$$e_1,\ e_2,\ -\frac{1}{2\mu}e_4,\ -\frac{1}{2\mu}e_3,$$

hence the trace of a symmetric 2-tensor T will be

$$\operatorname{tr} T = T(e_1, e_1) + T(e_2, e_2) - \frac{1}{\mu} T(e_3, e_4).$$

In general, for a given metric g, we wish to construct "good" null frames, that is, null frames which make the computations as easy as possible (in particular, we try to arrange $\mu = 1$ in most cases). There follow a few examples.

1. Quasiradial frame Let g be a metric on $\mathbf{R}^4_{x,t}$ satisfying $g^{00} = -1$, $\sum_i g^{0i}\omega^i = 0$, $\omega = x/r$. In particular, a split metric will satisfy the conditions. Set

$$T = -\nabla t = -g^{\alpha\beta}(\partial_\alpha t)\partial_\beta = -g^{0\beta}\partial_\beta = \partial_t - g^{0i}\partial_i.$$

We observe that $\langle T, T\rangle = \nabla t(t) = -1$, and that T is orthogonal to the surface

$$\Sigma_t = \{(x, t)\}.$$

We then define $N = \nabla r / |\nabla r|$, a unit vector orthogonal to the standard spheres

$$\{t = t_0,\ r = r_0\}.$$

Moreover, T and N are orthogonal, since $\partial_i r = x^i/r = \omega^i$ and

$$|\nabla r|\langle T, N\rangle = T(r) = -\sum_i g^{0i}\omega^i = 0.$$

Hence, if we take (e_1, e_2) to be an orthonormal basis (in the sense of g) on the standard spheres, the basis

$$e_1, e_2, e_3 = T - N, e_4 = T + N$$

is a null frame, called a quasiradial frame, with $\langle e_3, e_4\rangle = -2$. In the case of the Minkowski metric, $T = \partial_t$, $N = \partial_r$, and the construction yields the classical example.

The advantage of this choice is its explicit character and simplicity, since it involves only the foliation by the hypersurfaces Σ_t and the foliation by the standard 2-spheres: We will see that it is sufficient for many applications. Note that the null frame for the Schwarzschild metric given above is essentially a quasiradial frame (apart from the fact that g^{00} is not -1). It turns out, however, that there can be good reasons to introduce nonstandard spheres, as we shall see presently.

2. Null frame associated to one optical function Let $g = -dt^2 + g_{ij}dx^i dx^j$ be a split metric on $\mathbf{R}^4_{x,t}$ (close to the Minkowski metric) and u an optical function for g (close to $t - r$); this is, for instance, the framework used in [28]. Using the coordinate t, we define the foliation $\Sigma_{t_0} = \{(x,t), t = t_0\}$ as in **1**, and using u, we define the foliation by nonstandard 2-spheres as

$$S_{t_0,u_0} = \{(x,t),\ t = t_0,\ u(x,t) = u_0\}.$$

We then set $L = -\nabla u = (\partial_t u)\partial_t - (g^{ij}\partial_i u)\partial_j$. Since ∇u is orthogonal to $\{u = u_0\}$ and ∂_t is orthogonal to Σ_{t_0}, the field $\tilde{N} = -(g^{ij}\partial_i u)\partial_j$ is an horizontal field orthogonal to S_{t_0,u_0}. Moreover,

$$\langle \tilde{N}, \tilde{N}\rangle = g_{ij}(g^{ik}\partial_k u)(g^{jl}\partial_l u) = g^{kl}\partial_k u \partial_l u = (\partial_t u)^2.$$

We set $a = (\partial_t u)^{-1}$, $N = a\tilde{N}$. Then, if (e_1, e_2) form an orthonormal basis on the nonstandard spheres, the frame

$$e_1, e_2, e_3 \equiv \underline{L} = a(\partial_t - N), e_4 \equiv L = a^{-1}(\partial_t + N)$$

is a null frame with $\langle \underline{L}, L\rangle = -2$. Here again, in the case of the Minkowski metric, choosing $u = t - r$, we obtain $a = 1$, the spheres are the standard ones, and

$$L = -\nabla u = \partial_t + \partial_r,\ \underline{L} = \partial_t - \partial_r,$$

so that we obtain the standard example.

3. Null frame associated to two optical functions A more symmetric approach for a general metric g is to consider two optical functions u and $\underline{u}$, and define the sphere foliation by

$$S_{u_0,\underline{u}_0} = \{(x,t), u(x,t) = u_0,\ \underline{u}(x,t) = \underline{u}_0\}.$$

One can think of u and $\underline{u}$ as being close to $t - r$ and $t + r$, the spheres being nonstandard 2-spheres close to the usual ones. The advantage of doing so is that we do not have to consider any t-coordinate, which is more satifying in the context of relativity theory. We then set

$$L = -\nabla u,\ \underline{L} = -\nabla \underline{u},\ 2\Omega^2 = -\langle L, \underline{L}\rangle = -(g^{\alpha\beta}\partial_\alpha u \partial_\beta \underline{u})^{-1}.$$

The desired null frame is

$$e_1, e_2, e_3 = 2\Omega\underline{L},\ e_4 = 2\Omega L,\ \langle e_3, e_4\rangle = -2,$$

if, as before, (e_1, e_2) form an orthonormal basis of the spheres $S_{u_0,\underline{u}_0}$. As before, the standard example for the Minkowski metric is obtained by the choice $u = t - r$, $\underline{u} = t + r$.

4. Null frame associated to a sphere foliation More generally, following [29], we can start from a 2-spheres foliation chosen in such a way that the metric is positive definite on the tangent space to these 2-spheres. We then choose (e_1, e_2) to be an orthonormal basis on the spheres, and e_3 and e_4 to be independent null vectors in the orthogonal space to the spheres. The quasiradial case, for instance (the example in **1**), corresponds to choosing the standard spheres for this foliation, a simple choice which turns out to be sufficient for many applications. Working with nonstandard spheres as in **2** or **3** can be delicate. Note that for the example given above for the Kerr metrics, (e_1, e_2) are not tangent to a sphere foliation, since

$$[e_1, e_2] = (r^2 + a^2)\Sigma^{-\frac{3}{2}} \cos\theta[a\Sigma^{-\frac{1}{2}}(e_3 + e_4) - (\sin\theta)^{-1}e_2]$$

is not generated by (e_1, e_2).

In the construction of a null frame associated with a sphere foliation, e_4 is orthogonal to the planes generated by (e_1, e_2, e_4); if the distribution of these planes is integrable, that is, if there exists u such that these planes are tangent to the hypersurfaces $\{u = C\}$, then e_4 is colinear to ∇u, hence u is an optical function as in **2**. This shows how optical functions appear naturally in this framework. Note that quasiradial frames are not integrable in general.

3

Computing with frames

Performing computations in a variable frame very often requires taking the derivatives of components. A typical example is an expression of the form $X\langle Y, Z\rangle$, where (X, Y, Z) are vector fields: the result involves both derivatives of the coefficients of the metric and derivatives of the coefficients of the fields Y, Z. Even such simple computations quickly become impossible if one does not use the appropriate geometric tool, the metric connexion.

3.1 Metric connexion

1. A **connexion** is a derivation operator D of one vector field by another, yielding a new vector field

$$(X, Y) \mapsto D_X Y,$$

with the following properties:

(i) For any *function* $f \in C^\infty$, $D_{fX}Y = f D_X Y$. We say that D is "linear" in X; this implies in particular that $(D_X Y)(m)$ depends only on $X(m)$.

(ii) For any function f, $D_X(fY) = f D_X Y + (Xf)Y$. This is similar to the usual derivation of a product.

(iii) For any fields X, Y, $D_X Y - D_Y X = [X, Y]$, where $[X, Y] \equiv (XY - YX)$ is the bracket of the fields. We say that D is "torsion free".

We admit the fundamental theorem (see [22]), saying that there exists a unique **metric connexion** that is a connexion with the additional property

$$X\langle Y, Z\rangle = \langle D_X Y, Z\rangle + \langle Y, D_X Z\rangle.$$

This formula seems to ignore the derivatives of the coefficients of the metric, which are, however, present in $\langle Y, Z\rangle$. This is because these derivatives are, in fact, part of the definition of D, as we see now. In local coordinates, set

$$D_{\partial_\alpha} \equiv D_\alpha, \ D_{\partial_\alpha}\partial_\beta \equiv D_\alpha\partial_\beta = \Gamma^\gamma_{\alpha\beta}\partial_\gamma,$$

thus defining the **Christoffel symbols** $\Gamma^\gamma_{\alpha\beta}$. The torsion free character of D implies the symmetry $\Gamma^\gamma_{\alpha\beta} = \Gamma^\gamma_{\beta\alpha}$, since

$$\Gamma^\gamma_{\alpha\beta}\partial_\gamma = D_\alpha\partial_\beta = [\partial_\alpha, \partial_\beta] + D_\beta\partial_\alpha = \Gamma^\gamma_{\beta\alpha}\partial_\gamma.$$

Using the properties in the definition of D, it is easy to obtain the formula (note that we have lowered the first index!)

$$\Gamma_{\gamma\alpha\beta} \equiv g_{\gamma\nu}\Gamma^\nu_{\alpha\beta} = \langle D_\alpha\partial_\beta, \partial_\gamma\rangle = \tfrac{1}{2}(\partial_\alpha g_{\beta\gamma} + \partial_\beta g_{\alpha\gamma} - \partial_\gamma g_{\alpha\beta}).$$

In fact,

$$\begin{aligned}\langle D_\alpha\partial_\beta, \partial_\gamma\rangle &= \partial_\alpha g_{\beta\gamma} - \langle\partial_\beta, D_\alpha\partial_\gamma\rangle = \partial_\alpha g_{\beta\gamma} - \partial_\gamma g_{\alpha\beta} + \langle D_\gamma\partial_\beta, \partial_\alpha\rangle \\ &= \partial_\alpha g_{\beta\gamma} - \partial_\gamma g_{\alpha\beta} + \partial_\beta g_{\alpha\gamma} - \langle\partial_\gamma, D_\alpha\partial_\beta\rangle.\end{aligned}$$

The simplest example of connexion in $\mathbf{R}^4$ is

$$D_X(Y^\alpha\partial_\alpha) = X(Y^\alpha)\partial_\alpha,$$

which is the metric connexion corresponding to a constant coefficient metric. In general, using properties (i), (ii) and (iii), we have

$$D_X Y = X(Y^\beta)\partial_\beta + \Gamma^\gamma_{\alpha\beta}X^\alpha Y^\beta\partial_\gamma.$$

2. It is very useful to extend D to derive any tensor field T. The natural way to do this is to generalize the product formula; if T acts on p vectors, we set

$$\begin{aligned}X[T(Y_1, \ldots, Y_p)] = (D_X T)(Y_1, \ldots, Y_p) &+ T(D_X Y_1, Y_2, \ldots, Y_p) \\ &+ \cdots + T(Y_1, \ldots, D_X Y_p),\end{aligned}$$

and this formula defines $D_X T$. For instance,

$$X[g(Y_1, Y_2)] = (D_X g)(Y_1, Y_2) + g(D_X Y_1, Y_2) + g(Y_1, D_X Y_2),$$

which, by comparison with the formula defining a metric connexion, gives, $D_X g = 0$. This is a striking way of saying that the metric coefficients do not enter the formula for the derivative of a scalar product. Another instructive example is the computation of $D_X\omega$ for a 1-form ω. For any vector field Y,

$$X(\omega(Y)) = (D_X\omega)(Y) + \omega(D_X Y).$$

If Z is the vector field dual to ω, by the metric property,

$$\begin{aligned}\langle D_X Z, Y\rangle &= -\langle Z, D_X Y\rangle + X(\langle Z, Y\rangle) = -\omega(D_X Y) + X(\omega(Y)) \\ &= (D_X\omega)(Y).\end{aligned}$$

In other words, $D_X Z$ is also dual to $D_X \omega$: We said before we would not distinguish between ω and its dual Z. This holds also when we take derivatives: there is no need to know what object we derive (form or field), the result is the same.

3. The **divergence** of a vector field X is defined as

$$\operatorname{div} X = D_\alpha X^\alpha.$$

It is important to note that here and in what follows, $D_\alpha X^\alpha$ *never* means that we take the derivative ∂_α of the function X^α; it means that we compute *first* $D_\alpha X$, and *then* take the α-coordinate:

$$D_\alpha X^\alpha \equiv [D_\alpha X]^\alpha.$$

In local coordinates, using the above formula for the Christoffel symbols,

$$\operatorname{div} X = \partial_\alpha(X^\alpha) + X^\beta \Gamma^\alpha_{\alpha\beta} = \partial_\alpha(X^\alpha) + \tfrac{1}{2} g^{\alpha\beta} X(g_{\alpha\beta}).$$

Using $(\log\det A)' = \operatorname{tr}(A'A^{-1})$, we also obtain the useful formula

$$\partial_\alpha |g| = |g| g^{\beta\gamma} \partial_\alpha g_{\beta\gamma}, \ \operatorname{div} X = |g|^{-\frac{1}{2}} \partial_\alpha (X^\alpha |g|^{\frac{1}{2}}).$$

4. The following lemma is a consequence of $D_X g = 0$.

Lemma *(i) Let T be a 2-tensor and X any vector field. Then*

$$X(\operatorname{tr} T) \equiv X(T^\alpha_\alpha) = \operatorname{tr} D_X T \equiv D_X T^\alpha_\alpha.$$

(ii) Similarly, $X|T|^2 = 2 D_X T_{\alpha\beta} T^{\alpha\beta}$.

Proof To prove the first formula, we note that, in any frame (e_α), $\operatorname{tr} T = T(e_\alpha, e^\alpha)$. Hence

$$X(\operatorname{tr} T) = (D_X T)(e_\alpha, e^\alpha) + T(D_X e_\alpha, e^\alpha) + T(e_\alpha, D_X e^\alpha).$$

Since $\langle e^\alpha, e_\beta \rangle = \delta^\alpha_\beta$, $(D_X e^\alpha)_\beta = -(D_X e_\beta)^\alpha$. This implies that the last two terms in the formula for $X(\operatorname{tr} T)$ cancel out.

The proof of the second formula is similar:

$$\begin{aligned} X(T_{\alpha\beta} T^{\alpha\beta}) &= X(T_{\alpha\beta}) T^{\alpha\beta} + T_{\alpha\beta} X(T^{\alpha\beta}) \\ &= 2(D_X T)_{\alpha\beta} T^{\alpha\beta} + T^{\alpha\beta}[T(D_X e_\alpha, e_\beta) + T(e_\alpha, D_X e_\beta)] \\ &\quad + T_{\alpha\beta}[T(D_X e^\alpha, e^\beta) + T(e^\alpha, D_X e^\beta)]. \end{aligned}$$

Using $(D_X e^\alpha)_\beta = -(D_X e_\beta)^\alpha$ as before, we see that the T terms cancel out. ◇

3.2 Submanifolds

If $S \subset M$ is a submanifold of M, the restriction of g to vectors tangent to S gives a metric on S, called the **induced metric**. If S has codimension 1 with unit normal N, we define the bilinear **second form** k, acting on vector fields X, Y tangent to S, by

$$k(X, Y) = -\langle D_X N, Y \rangle.$$

Note that k is symmetric, since

$$\begin{aligned} k(X, Y) &= -X\langle Y, N\rangle + \langle N, D_X Y\rangle = \langle N, [X, Y] + D_Y X\rangle \\ &= \langle D_Y X, N\rangle = -\langle D_Y N, X\rangle = k(Y, X). \end{aligned}$$

We have used here the torsion free character of D, the metric property, and the fact that the Lie bracket $[X, Y]$ of X and Y is also tangent to S.

Example 1 Let $g = -dt^2 + g_{ij}dx^i dx^j$ be a split metric; the second form of $S = \Sigma_t$ is given by $k_{ij} = -\frac{1}{2}\partial_t g_{ij}$, since

$$N = \partial_t, \ \langle D_i \partial_t, \partial_j\rangle = \Gamma_{ji0} = \tfrac{1}{2}\partial_t g_{ij}.$$

Example 2 Consider in $\mathbf{R}^3_x$ the standard (flat) Riemannian metric and let S be the sphere of radius R in $\mathbf{R}^3$; then

$$N = R^{-1}x^i\partial_i, \ D_X N = R^{-1}X(x^i)\partial_i = R^{-1}X$$

and $k(X, Y) = -R^{-1}\langle X, Y\rangle$. Keep in mind, in particular, that the trace of k is $-2/R$.

For vectors X, Y tangent to the hypersurface S, we decompose $D_X Y$ into its tangential and normal parts:

$$D_X Y = T(X, Y) + R(X, Y).$$

Theorem *We have $R(X, Y) = k(X, Y)N$ and $T(X, Y)$ is the metric connexion $\not{D}_X Y$ on S associated to the induced metric. Hence*

$$D_X Y = \not{D}_X Y + k(X, Y)N.$$

Proof Since

$$\langle D_X Y, N\rangle = \langle R(X, Y), N\rangle = k(X, Y),$$

we have $R(X, Y) = k(X, Y)N$. It is easy to check that $T(X, Y)$ has properties (i), (ii) and (iii) of a connexion on S. To check the metric property, we observe that, for a field Z tangent to S,

$$Z\langle X, Y\rangle = \langle D_Z X, Y\rangle + \langle X, D_Z Y\rangle = \langle T(Z, X), Y\rangle + \langle X, T(Z, Y)\rangle.$$

Since the metric connexion is unique, T is *the* metric connexion on S, denoted by $\not{D}_X Y$. ◇

Finally, let us mention the **Stokes formula** in this context. Let $\mathcal{D}$ be an open domain (we assume that there can be no confusion between the domain and the connexion!) of $\mathbf{R}^4$ with smooth boundary $\partial\mathcal{D}$, and X be a vector field on $\mathcal{D}$. Then

$$\int_{\mathcal{D}} \operatorname{div} X dV = \int_{\partial\mathcal{D}} \langle X, N\rangle dv.$$

Here, the oriented unit normal N is defined by $N = \nabla f/||\nabla f||$, if f defines $\partial\mathcal{D}$ and $f < 0$ in $\mathcal{D}$; the volume elements on $\mathcal{D}$ and $\partial\mathcal{D}$ are respectively dV and dv. We refer the reader to Spivak [46] for details.

3.3 Hessian and d'Alembertian

1. For a given C^2 function f, we define the **Hessian** $\nabla^2 f$ of f as the bilinear form

$$\nabla^2 f(X, Y) = \langle D_X \nabla f, Y\rangle.$$

More explicitly,

$$\nabla^2 f(X, Y) = X\langle\nabla f, Y\rangle - \langle\nabla f, D_X Y\rangle = X(Y(f)) - (D_X Y)f.$$

This formula gives in particular

$$\begin{aligned}\nabla^2 f(X, Y) - \nabla^2 f(Y, X) &= (XY - YX)(f) - (D_X Y - D_Y X)(f)\\ &= [X, Y]f - [X, Y]f = 0,\end{aligned}$$

that is, $\nabla^2 f$ is bilinear *symmetric*. Note that, at a critical point m for f (that is, $\nabla f(m) = 0$),

$$\nabla^2 f(X, Y) = (\partial^2_{\alpha\beta} f)(m) X^\alpha Y^\beta.$$

The same formula holds everywhere for a constant coefficient metric, since then

$$\nabla^2 f X, Y) = X^\alpha Y^\beta \partial^2_{\alpha\beta} f + X^\alpha \partial_\alpha(Y^\beta)\partial_\beta f - X(Y^\beta)\partial_\beta f.$$

2. The **d'Alembertian** $\Box_g f = \Box f$ of f is defined as the trace of $\nabla^2 f$; from the formula in section 3.1 we get the various representations

$$\begin{aligned}\Box f &= (\nabla^2 f)^\alpha_\alpha = \langle D_\alpha \nabla f, \partial^\alpha\rangle = \operatorname{div}\nabla f = |g|^{-\frac{1}{2}}\partial_\alpha(g^{\alpha\beta}|g|^{\frac{1}{2}}\partial_\beta f),\\ \Box f &= g^{\alpha\beta}(\nabla^2 f)\langle\partial_\alpha, \partial_\beta\rangle = g^{\alpha\beta}[\partial^2_{\alpha\beta} f - (D_\alpha \partial_\beta) f] = \partial^\alpha\partial_\alpha f - (D^\alpha\partial_\alpha) f,\\ \Box f &= g^{\alpha\beta}\partial^2_{\alpha\beta} f + [\partial_\alpha(g^{\alpha\beta}) + \tfrac{1}{2} g^{\lambda\mu}\partial^\beta g_{\lambda\mu}]\partial_\beta f.\end{aligned}$$

Note that the principal symbol of $\Box$ is $p = g^{\alpha\beta}\xi_\alpha\xi_\beta$, but there are also lower order terms in $\Box$.

In a null frame (e_1, e_2, e_3, e_4), using the formula for the trace of a symmetric tensor, we get

$$\begin{aligned}\Box f &= -\nabla^2 f(e_3, e_4) + \nabla^2 f(e_1, e_1) + \nabla^2 f(e_2, e_2) \\ &= -e_4 e_3 f + (e_1^2 + e_2^2) f + [D_4 e_3 - (D_1 e_1 + D_2 e_2)] f.\end{aligned}$$

One has to be careful in interpreting this formula, as we can see in the flat case with the usual d'Alembertian. Then,

$$e_4 e_3 = \partial_t^2 - \partial_r^2, \; D_4 e_3 = 0.$$

Using spherical coordinates and taking $e_1 = r^{-1}\partial_\theta$, $e_2 = (r \sin\theta)^{-1}\partial_\phi$, we have

$$D_1 e_1 = -r^{-1}\partial_r, \; D_2 e_2 = -(r \sin\theta)^{-2}(x^1\partial_1 + x^2\partial_2).$$

Hence, using the definition of the induced connexion on the sphere,

$$\not{D}_1 e_1 = 0, \; D_2 e_2 = -r^{-1}\partial_r + \not{D}_2 e_2, \; \not{D}_2 e_2 = -r^{-2}\frac{\cos\theta}{\sin\theta}\partial_\theta.$$

Gathering the terms,

$$D_1 e_1 + D_2 e_2 = -\frac{2}{r}\partial_r - r^{-2}\frac{\cos\theta}{\sin\theta}\partial_\theta.$$

Finally, we get

$$\Box = -\partial_t^2 + \partial_r^2 + \frac{2}{r}\partial_r + \frac{1}{r^2}\Delta_S, \; \Delta_S = \partial_\theta^2 + (\sin\theta)^{-2}\partial_\phi^2 + \frac{\cos\theta}{\sin\theta}\partial_\theta.$$

We recognize the usual expression of the d'Alembertian in spherical coordinates, Δ_S being the Laplacian on the unit sphere (see [9] for details).

3. Geodesics, bicharacteristics and optical functions

a. *Geodesics* Let $x(s)$ be a C^2 curve defined on some interval I. We say that this curve is a **geodesic** if its tangent vector $T = (d/ds)x$ satisfies $D_T T = 0$. Note that this makes sense though T is not defined everywhere; note also that this condition depends on the way the curve is parametrized. Denoting by a dot the s-derivative, and observing that $(d/ds)(f(x(s))) = Tf(x(s))$,

$$D_T T = D_T \dot{x} = T(\dot{x}^\alpha)\partial_\alpha + \Gamma^\alpha_{\beta\gamma}\dot{x}^\beta\dot{x}^\gamma\partial_\alpha.$$

Hence a geodesic is a curve satisfying the set of differential equations

$$\frac{d^2}{ds^2}x^\alpha + \Gamma^\alpha_{\beta\gamma}\left(\frac{d}{ds}x^\beta\right)\left(\frac{d}{ds}x^\gamma\right) = 0.$$

For instance, assume u to be an optical function, and set $L = -\nabla u$. Then $D_L L = 0$, showing that an integral curve of L (that is, $\dot{x} = L$) is a geodesic. This follows from the symmetry of the Hessian, since for any X

$$\begin{aligned}\langle D_L L, X\rangle &= -\langle D_L \nabla u, X\rangle = -\langle D_X \nabla u, L\rangle = \langle D_X L, L\rangle \\ &= \tfrac{1}{2} X\langle L, L\rangle = 0.\end{aligned}$$

b. ***Bicharacteristics*** To the metric g is associated the function $p(x, \xi) = g^{\alpha\beta}\xi_\alpha \xi_\beta$. This is a well-defined function on the cotangent space of the manifold M, which is the principal symbol of the wave operator $\Box$. From a PDE point of view, it is important to consider (null) bicharacteristic curves of $\Box$ starting from a (characteristic) point (x_0, ξ^0), which are defined by

$$\frac{d}{ds}x^\alpha \equiv \dot{x}^\alpha = \partial_{\xi_\alpha} p,\ \dot{\xi}_\alpha = -\partial_\alpha p,\ x^\alpha(0) = x_0^\alpha,\ \xi_\alpha(0) = \xi_\alpha^0,\ p(x_0, \xi^0) = 0.$$

If we start from the point $(x_0, \mu\xi^0)$, the solution is just $(x(\mu s), \mu\xi(\mu s))$. Differentiating the system once more, we get an autonomous differential equation for the coordinates x^α:

$$\begin{aligned}\frac{d^2}{ds^2}x^\alpha &= \partial_\gamma g^{\alpha\beta} g_{\beta\mu}\dot{x}^\gamma \dot{x}^\mu - \frac{1}{2}\partial^\alpha g^{\lambda\mu} g_{\lambda\lambda'} g_{\mu\mu'}\dot{x}^{\lambda'}\dot{x}^{\mu'} \\ &= -g^{\alpha\beta}\partial_\gamma g_{\beta\mu}\dot{x}^\gamma \dot{x}^\mu + \frac{1}{2}\partial^\alpha g_{\lambda\lambda'}\dot{x}^\lambda \dot{x}^{\lambda'} = -\Gamma^\alpha_{\beta\gamma}\dot{x}^\beta \dot{x}^\gamma.\end{aligned}$$

We recognize the equation of a **geodesic curve**. The initial conditions for the geodesic curve, which is the projection of the bicharacteristic, are

$$x(0) = x_0,\ \dot{x}(0)^\alpha = 2g^{\alpha\beta}\xi^0_\beta.$$

c. ***Geodesic cone*** For a given point x_0, consider a nonzero future oriented null vector ξ^0 at this point. There is a unique bicharacteristic curve starting from (x_0, ξ^0); the union of all such half-curves (for $s \geq 0$) starting from x_0 forms the (half) geodesic cone with summit at x_0. Let L be the vector field $\dot{x}$ on this cone. For each value of the parameter s, let S_s be the locus of the points $x(s)$ for the various ξ_0. Choose a nonzero field X tangent to, say, S_{s_0}, and define X on the cone to be X extended by the action of the flow of L.

Lemma *The vector L is a null vector that is orthogonal to the geodesic cone.*

Proof First, $\langle L, L\rangle = 0$, since $L\langle L, L\rangle = 2\langle D_L L, L\rangle = 0$ and, for $s = 0$,

$$\langle L, L\rangle = g_{\alpha\beta}\dot{x}^\alpha \dot{x}^\beta = 4p(x_0, \xi^0) = 0.$$

Next, $[L, X] = 0$ by construction. Then, for the induced connexion $\not{D}$ on the cone,

$$L\langle X, L\rangle = \langle \not{D}_L X, L\rangle + \langle X, \not{D}_L L\rangle$$
$$= \langle [L, X], L\rangle + \langle \not{D}_X L, L\rangle = \tfrac{1}{2} X\langle L, L\rangle = 0.$$

Since $\langle X, L\rangle$ goes to zero when s goes to zero, $\langle X, L\rangle = 0$ and the orthogonal to L is the tangent plane to the cone. ◇

Consider now a one-parameter family of geodesic cones such that there exists a function u that has the cones of this family as level sets. A typical case would, for the Minkowski metric, be the geodesic cones with vertices on the t-axis, corresponding to a function $u = F(t - r)$ (which is, of course, singular on the t-axis). Since ∇u is then normal to each geodesic cone, ∇u is colinear to L, hence ∇u is a null vector and u an optical function.

d. *Optical functions* This construction can be extended if, instead of starting from a single point x_0, we start from a spacelike 2-surface S_0. Choosing at each point x_0 of S_0 an outgoing future oriented null vector $\xi^0(x_0)$ orthogonal to S_0, we consider the union Σ of all the geodesic curves issued from $(x_0, \xi^0(x_0))$. Defining L and X on Σ as before, we obtain $L\langle X, L\rangle = 0$ as before, hence $\langle X, L\rangle = 0$ since this is true by construction for $s = 0$. Again, if we are given a one-parameter family of such surfaces Σ, such that there exists a function u having these Σ as level sets, then u is an optical function.

3.4 Frame coefficients

As explained above, working in a given frame (e_α) requires that we know the vectors $D_\alpha e_\beta$, that is, the frame coefficients $\langle D_\alpha e_\beta, e_\gamma\rangle$. In the case of local coordinates x^α, $e_\alpha = \partial_\alpha$, $D_\alpha \partial_\beta = \Gamma^\gamma_{\alpha\beta}\partial_\gamma$ and we have seen already the explicit formula for Γ.

1. Computing the frame coefficients from the brackets In the case of local coordinates where $e_\alpha = \partial_\alpha$, the brackets $[e_\alpha, e_\beta]$ vanish, and the explicit formula for Γ was obtained using the basic properties of the connexion. More generally, we can do the same thing if we work with a null frame (e_α) for which we know the brackets $[e_\alpha, e_\beta]$. Using the various properties of the metric connexion, we can compute explicitly the vectors $D_\alpha e_\beta$. We give here a few examples of these manipulations:

(i) $\langle D_1 e_1, e_1\rangle = 0$, since $e_1\langle e_1, e_1\rangle = 0 = 2\langle D_1 e_1, e_1\rangle$;
(ii) $\langle D_1 e_1, e_2\rangle = -\langle e_1, D_1 e_2\rangle = -\langle e_1, D_2 e_1 + [e_1, e_2]\rangle = -\langle e_1, [e_1, e_2]\rangle$;

(iii) for $j = 3, 4$,

$$\langle D_j e_1, e_2\rangle = \langle [e_j, e_1], e_2\rangle + \langle D_1 e_j, e_2\rangle,$$
$$\langle D_1 e_j, e_2\rangle = -\langle e_j, D_1 e_2\rangle = -\langle e_j, [e_1, e_2]\rangle - \langle e_j, D_2 e_1\rangle,$$
$$\langle e_j, D_2 e_1\rangle = -\langle e_1, D_2 e_j\rangle = -\langle e_1, [e_2, e_j]\rangle + \langle e_2, D_j e_1\rangle.$$

Hence finally,

$$2\langle D_j e_1, e_2\rangle = \langle [e_j, e_1], e_2\rangle - \langle e_j, [e_1, e_2]\rangle + \langle e_1, [e_2, e_j]\rangle.$$

Example Let us compute, for instance, the frame coefficients for the quasiradial null frame of a quasiradial situation, as described in section 2.3. We define the second form of the foliation $\Sigma_T = \{(x, t), t = T\}$ by $k(X, Y) = -\langle D_X T, Y\rangle$. In local coordinates,

$$\begin{aligned} k_{ij} &= -\langle D_i T, \partial_j\rangle = (\nabla^2 t)_{ij} = -(D_i \partial_j)(t) = -\Gamma^0_{ij} \\ &= -\tfrac{1}{2} g^{0\alpha}(\partial_i g_{\alpha j} + \partial_j g_{\alpha i} - \partial_\alpha g_{ij}). \end{aligned}$$

For vector fields X, Y, we define $\not{D}_X Y$ to be the orthogonal projection of $D_X Y$ onto the space generated by (e_1, e_2)

$$\not{D}_X Y = \langle D_X Y, e_1\rangle e_1 + \langle D_X Y, e_2\rangle e_2.$$

Note that this definition extends that of the induced connexion on the spheres, since $\not{D}_X Y$ is the result of the induced connexion if X, Y are tangent to the sphere. In the sequence, indices a, b run from 1 to 2, corresponding to the basis on the spheres. We set

$$c = |\nabla r|,\ c^2 = \langle \nabla r, \nabla r\rangle = g^{ij}\omega_i\omega_j.$$

Theorem (Quasiradial case) *The connexion D satisfies*

$$D_T T = 0,\ D_N T = -k_{NN} N - k_{aN} e_a,\ D_a T = -k_{aN} N - k_{ab} e_b,$$
$$D_T N = k_{aN} e_a,\ D_N N = -k_{NN} T + \frac{e_a(c)}{c} e_a,\ D_a N = \not{D}_a N - k_{aN} T,$$
$$D_N e_a = \not{D}_N e_a - \frac{e_a(c)}{c} N - k_{aN} T,\ D_T e_a = \not{D}_T e_a - k_{aN} N.$$

Proof (a) For any field X, by symmetry of the Hessian, $\langle D_T T, X\rangle = \langle D_X T, T\rangle = 0$, hence $D_T T = 0$. For tangential X, Y, $\langle D_X T, T\rangle = 0$, $\langle D_X T, Y\rangle = -k(X, Y)$. This gives the first line.

(b) Since $g^{0i}\omega_i = 0$, using the formula for the decomposition of ∂_i into its radial part and its rotation part

$$\partial_i = \omega_i \partial_r - \left(\omega \wedge \frac{R}{r}\right)_i,$$

we obtain

$$T = \partial_t - g^{0i}(\partial_i - \omega_i \partial_r) = \partial_t + g^{0i}\left(\omega \wedge \frac{R}{r}\right)_i,$$

$$N = c^{-1} g^{ij} \omega_i \partial_j = c\partial_r - c^{-1} g^{ij} \omega_i \left(\omega \wedge \frac{R}{r}\right)_j.$$

This implies

$$[T, N] = \frac{Tc}{c} N + \cdots R.$$

We compute now the derivatives of $N = c^{-1}\nabla r$. First, $\langle D_T N, N\rangle = 0$,

$$\langle D_T N, T\rangle = \langle [T, N], T\rangle + \langle D_N T, T\rangle = 0.$$

Next,

$$\begin{aligned}\langle D_T N, e_a\rangle &= c^{-1}\langle D_T \nabla r, e_a\rangle = c^{-1}\langle D_a \nabla r, T\rangle = \langle D_a N, T\rangle \\ &= -\langle D_a T, N\rangle = k_{aN}.\end{aligned}$$

(c) Similarly, we know $D_N N$, since $\langle D_N N, N\rangle = 0$,

$$\langle D_N N, T\rangle = -\langle D_N T, N\rangle = k_{NN},$$

$$\langle D_N N, e_a\rangle = c^{-1}\langle D_N \nabla r, e_a\rangle = c^{-1}\langle D_a \nabla r, N\rangle = \frac{e_a(c)}{c}.$$

(d) Finally, $\langle D_a N, N\rangle = 0$,

$$\langle D_a N, T\rangle = c^{-1}\langle D_a \nabla r, T\rangle = c^{-1}\langle D_T \nabla r, e_a\rangle = \langle D_T N, e_a\rangle = k_{aN}.$$

The quantities $\langle D_a N, e_b\rangle = \langle \not{D}_a N, e_b\rangle$ are the components of the second form of the standard spheres with respect to the induced metric g_{ij} in Σ_t.

For any field X, $\langle D_X e_a, T\rangle = -\langle D_X T, e_a\rangle$ and $\langle D_X e_a, N\rangle = -\langle D_X N, e_a\rangle$ are already known. The other quantities $\langle D_N e_a, e_b\rangle$, $\langle D_T e_a, e_b\rangle$ and $\langle D_c e_a, e_b\rangle$ depend on the choice of (e_1, e_2), and have to be computed using the expressions of the brackets as above. ◇

2. Intrinsic frame coefficients In the case of a frame associated to one or two optical functions (see section 2.3), using the properties of the null frame and of the connexion, one can deduce all frame coefficients from the following special frame coefficients:

(i) Define first the analogs of the second form for the nonstandard spheres by

$$\chi_{ab} = \langle D_a L, e_b\rangle,\ \underline{\chi}_{ab} = \langle D_a \underline{L}, e_b\rangle,\ \underline{L} = e_3,\ L = e_4,\ a, b = 1, 2.$$

These two 2-tensors are symmetric for the same reason as for the second form of a hypersurface, namely, because $[e_a, e_b]$ is tangent to the spheres.

(ii) Next, define four 1-forms on the spheres by

$$2\eta_a = \langle D_{\underline{L}} L, e_a \rangle, \; 2\underline{\eta}_a = \langle D_L \underline{L}, e_a \rangle,$$
$$2\xi_a = \langle D_L L, e_a \rangle, \; 2\underline{\xi}_a = \langle D_{\underline{L}} \underline{L}, e_a \rangle.$$

(iii) Finally, define the functions ω and $\underline{\omega}$ by

$$4\omega = \langle D_L L, \underline{L} \rangle, \; 4\underline{\omega} = \langle D_{\underline{L}} \underline{L}, L \rangle.$$

We check now that these quantities allow us to recover all the frame coefficients. Suppose the frame is associated to one optical function u, with $L = -\nabla u$ as explained in the second example of section 2.3.

Theorem (Integrable case) *The connexion D satisfies the formulas*

$$D_L L = 0, \; D_{\underline{L}} L = 2\eta_a e_a + 2\underline{\omega} L, \; D_a L = \chi_{ab} e_b - \eta_a L,$$
$$D_L \underline{L} = 2\underline{\eta}_a e_a, \; D_{\underline{L}} \underline{L} = 2\underline{\xi}_a e_a - 2\underline{\omega} \underline{L}, \; D_a \underline{L} = \underline{\chi}_{ab} e_b + \eta_a \underline{L},$$
$$D_L e_a = \not{D}_L e_a + \underline{\eta}_a L, \; D_{\underline{L}} e_a = \not{D}_{\underline{L}} e_a + \underline{\xi}_a L + \eta_a \underline{L},$$
$$D_b e_a = \not{D}_b e_a + \tfrac{1}{2}\chi_{ab} \underline{L} + \tfrac{1}{2}\underline{\chi}_{ab} L,$$
$$\eta_a = \frac{e_a(a)}{a} + k_{aN}.$$

Proof First, we note that $D_L L = 0$, since

$$\langle D_L L, X \rangle = -\langle D_L \nabla u, X \rangle = -\nabla^2 u(L, X)$$
$$= -\nabla^2 u(X, L) = -\langle D_X \nabla u, L \rangle = \langle D_X L, L \rangle = 0.$$

This means that the integral curves of L along the outgoing cones are geodesics. Also, since $T \equiv \partial_t = -\nabla t$, we get $\langle D_T T, X \rangle = \langle D_X T, T \rangle = 0$, which shows that $D_T T = 0$. All other formulas are easy to prove, except the last one; for this, we compute $e_a(\langle L, T \rangle)$ in two different ways:

$$\langle L, T \rangle = -\frac{1}{a}, \; e_a(\langle L, T \rangle) = \frac{e_a(a)}{a^2},$$
$$e_a(\langle L, T \rangle) = \langle D_a L, T \rangle + \langle L, D_a T \rangle.$$

Now $\underline{L} = -a^2 L + 2aT$, which gives

$$\langle D_a L, T \rangle = (2a)^{-1} \langle D_a L, \underline{L} + a^2 L \rangle = \frac{\eta_a}{a}.$$

Also,

$$\langle L, D_a T \rangle = a^{-1} \langle T + N, D_a T \rangle = -a^{-1} k_{aN},$$

which proves the formula. ◇

We prove now that the underlined quantities can be recovered from the others and from the second form $k(X, Y) = -\langle D_X \partial_t, Y \rangle$.

Theorem *The following formulas hold*

$$\underline{\chi}_{ab} = -a^2\chi_{ab} - 2ak_{ab},\ \underline{\xi}_a = -a^2\eta_a + a^2k_{aN},$$
$$\underline{\eta}_a = -k_{aN},\ 2\underline{\omega} = ak_{NN} - \frac{\underline{L}a}{a}.$$

Proof For instance, since $\underline{L} = -a^2L + 2a\partial_t$,

$$\underline{\chi}_{ab} = -a^2\chi_{ab} - 2ak_{ab},$$
$$2\underline{\xi}_a = \langle D_{\underline{L}}\underline{L}, e_a\rangle = -2a^2\eta_a + 2a\langle D_{\underline{L}}\partial_t, e_a\rangle,$$
$$\langle D_{\underline{L}}\partial_t, e_a\rangle = \langle D_a\partial_t, \underline{L}\rangle = -a\langle D_a\partial_t, N\rangle = -a\langle D_N\partial_t, e_a\rangle = ak_{aN},$$
$$2\underline{\eta}_a = \langle D_L(-a^2L + 2aT), e_a\rangle = 2a\langle D_L\partial_t, e_a\rangle = 2\langle D_NT, e_a\rangle = -2k_{aN}.$$

Finally, we check $2\underline{\omega} = ak_{NN} - \underline{L}a/a$. In fact,

$$\begin{aligned}\langle D_{\underline{L}}L, \underline{L}\rangle &= a^2\langle D_{T-N}(a^{-1}(T+N)), T-N\rangle\\ &= a^2[-2(T-N)(a^{-1}) + a^{-1}\langle D_{T-N}(T+N), T-N\rangle],\\ &\langle D_{T-N}(T+N), T-N\rangle = -2k_{NN}.\end{aligned}$$

◇

To conclude, note that the quantities χ, η are components of the Hessian ∇^2u:

$$\chi_{ab} = -\nabla^2u_{ab},\ 2\eta_a = -\nabla^2u_{a\underline{L}}.$$

Note in particular that, according to the trace formula of chapter 1,

$$\Box u = \operatorname{tr}\nabla^2u = \nabla^2u_a^a - \nabla^2u_{34} = -\operatorname{tr}\chi + \langle D_{\underline{L}}L, L\rangle = -\operatorname{tr}\chi.$$

In chapter 6, we will discuss in more detail how the tensors χ, ξ, etc. can be estimated in situations where they are not explicitly known.

4

Energy inequalities and frames

To obtain an energy inequality for the wave operator $\Box = \Box_g$ associated to a given Lorentzian metric g, we proceed as usual by choosing a vector field X (the "multiplier") and writing

$$(\Box\phi)(X\phi) = \operatorname{div} P + q.$$

Here, P will be an appropriate field whose coefficients are quadratic forms in the components of $\nabla\phi$, while q is a quadratic form in these components with variable coefficients. Integrating $(\Box\phi)(X\phi)$ in some domain $\mathcal{D}$, and using the Stokes formula, we obtain boundary terms

$$\int_{\partial\mathcal{D}} \langle P, N\rangle dv,$$

which yield the "energy" of ϕ, and interior terms $\int_{\mathcal{D}} q\,dV$. In practice, since some derivatives of ϕ behave better than others, we must write these energy and interior terms in an appropriate frame, and *not* in the usual coordinates. We describe now the clever machinery which makes this possible and easy.

4.1 The energy–momentum tensor

1. Let ϕ be a fixed C^1 function, and define the **energy–momentum tensor** Q as a symmetric 2-tensor by

$$Q(X,Y) = (X\phi)(Y\phi) - \tfrac{1}{2}\langle X, Y\rangle|\nabla\phi|^2, \; Q_{\alpha\beta} = (\partial_\alpha\phi)(\partial_\beta\phi) - \tfrac{1}{2}g_{\alpha\beta}|\nabla\phi|^2.$$

Consider a null frame (e_1, e_2, e_3, e_4), with $\langle e_3, e_4\rangle = -2\mu$. Since

$$\nabla\phi = e_1(\phi)e_1 + e_2(\phi)e_2 - \frac{1}{2\mu}(e_4(\phi)e_3 + e_3(\phi)e_4),$$

we get

$$|\nabla\phi|^2 = \langle\nabla\phi, \nabla\phi\rangle = e_1(\phi)^2 + e_2(\phi)^2 - \frac{1}{\mu}e_3(\phi)e_4(\phi).$$

Note that $|\nabla\phi|^2$ is not a positive term! Hence, writing for simplicity $e_\alpha(\phi) = e_\alpha$, the components $Q_{\alpha\beta} = Q(e_\alpha, e_\beta)$ of Q are

$$Q_{11} = \frac{1}{2}(e_1^2 - e_2^2) + \frac{1}{2\mu}e_3e_4,\ Q_{12} = e_1e_2,\ Q_{22} = -\frac{1}{2}(e_1^2 - e_2^2) + \frac{1}{2\mu}e_3e_4,$$
$$Q_{13} = e_1e_3,\ Q_{14} = e_1e_4,\ Q_{23} = e_2e_3,\ Q_{24} = e_2e_4,$$
$$Q_{33} = e_3^2,\ Q_{34} = \mu(e_1^2 + e_2^2),\ Q_{44} = e_4^2.$$

In particular, according to the formula for the trace given in section 3.3, the trace $\operatorname{tr} Q = Q_\alpha^\alpha$ is

$$\operatorname{tr} Q = Q_{11} + Q_{22} - \frac{1}{\mu}Q_{34} = -(e_1^2 + e_2^2) + \frac{1}{\mu}e_3e_4 = -|\nabla\phi|^2.$$

2. Let us say a few words about the geometry of g here. First, recall the terminology for vectors: a vector X is **timelike**, **null**, or **spacelike** if $\langle X, X\rangle < 0$, $\langle X, X\rangle = 0$ or $\langle X, X\rangle > 0$ respectively. A vector X is future oriented (resp. past oriented) if its t-component is positive (resp. negative). We have seen that in a null frame, the metric can be written

$$\langle X, X\rangle = (X^1)^2 + (X^2)^2 - 4\mu X^3X^4.$$

By a linear change of variables $X \mapsto Y$, this can be reduced to the canonical form $(Y^1)^2 + (Y^2)^2 + (Y^3)^2 - (Y^4)^2$. Hence the future oriented half-cone $\langle X, X\rangle < 0$ is a convex set, and the scalar product of two future oriented null vectors is nonpositive.

3. The tensor Q exhibits a remarkable positivity property.

Theorem (Positivity of Q) *If X and Y are two timelike future oriented vectors, then*

$$Q(X, Y) \geq 0.$$

Proof Let e_3 and e_4 be two *null* future oriented vectors in the plane generated by (X, Y). Then, by convexity of the future timelike cone,

$$X = ae_3 + a'e_4,\ Y = be_3 + b'e_4,\ a, a', b, b' \geq 0.$$

Hence, setting $\langle e_3, e_4\rangle = -2\mu,\ \mu \geq 0$,

$$Q(X, Y) = abe_3(\phi)^2 + a'b'e_4(\phi)^2 + \mu(ab' + a'b)[e_1(\phi)^2 + e_2(\phi)^2] \geq 0. \quad \diamond$$

4. The energy–momentum tensor and the d'Alembertian

Theorem *The energy-momentum tensor Q is related to the d'Alembertian by the formula*

$$D^\alpha Q_{\alpha\beta} = (\Box\phi)(\partial_\beta\phi).$$

Proof We write, using $D_X g = 0$,

$$Q = d\phi \otimes d\phi - \tfrac{1}{2} g |\nabla\phi|^2,$$
$$D_X Q = D_X d\phi \otimes d\phi + d\phi \otimes D_X d\phi - g \langle D_X \nabla\phi, \nabla\phi\rangle.$$

Taking $X = \partial^\alpha$, and then the $\alpha\beta$ component, we obtain

$$D^\alpha Q_{\alpha\beta} = (\nabla^2 \phi^\alpha_\alpha)(\partial_\beta\phi) + (\partial_\alpha\phi)\nabla^2\phi^\alpha_\beta - \nabla^2\phi(\partial_\beta, \nabla\phi).$$

On the right-hand side, the first term is by definition $(\Box\phi)(\partial_\beta\phi)$ and the second is $\nabla^2\phi(\nabla\phi, \partial_\beta)$, which cancels out with the third. ◇

4.2 Deformation tensor

Definition The deformation tensor of a given vector field X is the symmetric 2-tensor ${}^{(X)}\pi$ defined by

$${}^{(X)}\pi(Y, Z) \equiv \pi(Y, Z) = \langle D_Y X, Z\rangle + \langle D_Z X, Y\rangle.$$

In local coordinates (be careful about the position of the indices!),

$$\pi_{\alpha\beta} = D_\alpha X_\beta + D_\beta X_\alpha.$$

If we were in $\mathbf{R}^3$ with the standard scalar product and the trivial connexion, the deformation tensor ${}^{(X)}\pi$ would be simply represented by twice the symmetric part of X' (the differential of X).

1. Let us digress shortly to explain a few things about **Lie derivatives**. For a given field X, we define $\mathcal{L}_X f = Xf$, $\mathcal{L}_X Y = [X, Y]$, and extend this to tensors by imitating the product formula, exactly as we have done for D_X:

$$X(T(Y_1, \ldots, Y_p)) = (\mathcal{L}_X T)(Y_1, \ldots, Y_p) + \sum_{1\le i\le p} T(Y_1, \ldots, [X, Y_i], \ldots, Y_p).$$

As is well known, the Lie derivative is defined using transportation along the flow of X, and $\mathcal{L}_X Y$ is *not* linear in X, in contrast with the covariant derivative

$D_X Y$. Using this definition with $T = g$, we find

$$X(g(Y,Z)) = (\mathcal{L}_X g)(Y,Z) + g([X,Y],Z) + g(Y,[X,Z])$$
$$= g(D_X Y, Z) + g(Y, D_X Z),$$

and hence finally the important formula

$$\pi = \mathcal{L}_X g.$$

This formula helps one to visualize what π is; in particular, π vanishes if g is invariant by the flow of X. We call such a field a **Killing field**. If only $\mathcal{L}_X g = \lambda g$, the field is **conformal Killing**. For the Minkowski metric, the simplest examples of Killing fields are the derivations, the spatial rotations and the hyperbolic rotations $H_i = t\partial_i + x^i \partial_t$; note that among these fields, only ∂_t is timelike. There are five conformal Killing fields (again, be aware of the position of the index μ)

$$S = x^\alpha \partial_\alpha, \; K_\mu = -2x_\mu S + |x|^2 \partial_\mu,$$

for which the corresponding deformation tensors are

$${}^{(S)}\pi = 2g, \; {}^{(K_\mu)}\pi = -4x_\mu g.$$

Among these, only $K_0 = (r^2 + t^2)\partial_t + 2rt\partial_r$ is timelike. For the Schwarzschild metric, ∂_t and the spatial rotations are Killing fields; for the Kerr metric, ∂_t and ∂_ϕ are Killing. For a general metric, there are no Killing fields, because the number of equations to be satisfied is ten, while there are only four unknowns.

2. To compute ${}^{(X)}\pi$ explicitly in a frame, we need the frame coefficients. However, in local coordinates, we have the simple formula

$${}^{(X)}\pi^{\alpha\beta} = \partial^\alpha(X^\beta) + \partial^\beta(X^\alpha) - X(g^{\alpha\beta}).$$

In fact, $D_\alpha X_\beta = \langle D_\alpha X, \partial_\beta \rangle = g_{\beta\gamma}\partial_\alpha(X^\gamma) + X^\gamma \Gamma_{\beta\alpha\gamma}$. Hence

$$\pi^{\alpha\beta} = g^{\alpha\alpha'} g^{\beta\beta'} [g_{\beta'\gamma}\partial_{\alpha'}(X^\gamma) + X^\gamma \Gamma_{\beta'\alpha'\gamma} + g_{\alpha'\gamma}\partial_{\beta'}(X^\gamma) + X^\gamma \Gamma_{\alpha'\beta'\gamma}]$$
$$= \partial^\alpha(X^\beta) + \partial^\beta(X^\alpha) + X^\gamma g^{\alpha\alpha'} g^{\beta\beta'} [\Gamma_{\beta'\alpha'\gamma} + \Gamma_{\alpha'\beta'\gamma}].$$

Since, from the explicit formula, $\Gamma_{\alpha\beta\gamma} + \Gamma_{\beta\alpha\gamma} = \partial_\gamma(g_{\alpha\beta})$, the last term is

$$g^{\alpha\alpha'} g^{\beta\beta'} X(g_{\alpha'\beta'}) = -X(g^{\alpha\beta}).$$

3. Finally, let us keep in mind the following formula, which will be useful later:

$${}^{(fX)}\pi(Y,Z) = f\,{}^{(X)}\pi(Y,Z) + (Yf)\langle X, Z\rangle + (Zf)\langle X, Y\rangle,$$

where $f \in C^1$ is an arbitrary function. We will see that deformation tensors play a crucial role in both energy computations and commutation formulas.

4.3 Energy inequality formalism

Theorem (Key formula) *Let ϕ be a given C^2 function and Q be the associated energy–momentum tensor. Let X be a vector field, and set $P_\alpha = Q_{\alpha\beta} X^\beta$. Then*

$$\operatorname{div} P \equiv D_\alpha P^\alpha = (\Box\phi)(X\phi) + \tfrac{1}{2} Q^{\alpha\beta(X)}\pi_{\alpha\beta}.$$

Proof Using the definitions, we write

$$\begin{aligned} \partial^\alpha(P_\alpha) &= \partial^\alpha(P(\partial_\alpha)) = D^\alpha P_\alpha + P(D^\alpha \partial_\alpha) \\ &= \partial^\alpha(Q(\partial_\alpha, X)) = (D^\alpha Q)(\partial_\alpha, X) + Q(D^\alpha \partial_\alpha, X) + Q(\partial_\alpha, D^\alpha X). \end{aligned}$$

The terms $P(D^\alpha \partial_\alpha)$ and $Q(D^\alpha \partial_\alpha, X)$ cancel out, so that we are left with

$$D^\alpha P_\alpha = D^\alpha Q_{\alpha\beta} X^\beta + Q(\partial^\alpha, D_\alpha X).$$

Using the above formula for $D^\alpha Q_{\alpha\beta}$, we see that the first term in the right-hand side is $(\Box\phi)(X\phi)$; to deal with the second term, we note that

$$Q^{\alpha\gamma}\langle D_\alpha X, \partial_\gamma\rangle = \tfrac{1}{2} Q^{\alpha\gamma}[\langle D_\alpha X, \partial_\gamma\rangle + \langle D_\gamma X, \partial_\alpha\rangle] = Q\pi,$$

while $Q(\partial_\alpha, D^\alpha X) = Q(\partial^\alpha, D_\alpha X) = Q^{\alpha\gamma}\langle D_\alpha X, \partial_\gamma\rangle$. ◇

The formula in the theorem is the key formula for proving an energy inequality. Let us explain why. The formula is just a rewriting of $(\Box\phi)(X\phi)$ as the sum of the divergence of a field (the term $\operatorname{div} P$) and a quadratic form q in the components of $\nabla\phi$, as expected. The point is that q is written as a double trace (summation on α and β), which can be computed using any null frame. More explicitly, we have for any null frame (e_α), denoting as usual $e_\alpha(\phi) = e_\alpha$,

$$\begin{aligned} Q\pi = Q_{\alpha\beta}\pi^{\alpha\beta} &= Q_{11}\pi_{11} + 2\pi_{12}e_1e_2 - \frac{1}{\mu}\pi_{14}e_1e_3 - \frac{1}{\mu}\pi_{13}e_1e_4 + \pi_{22}Q_{22} \\ &\quad - \frac{1}{\mu}\pi_{24}e_2e_3 - \frac{1}{\mu}\pi_{23}e_2e_4 + \frac{1}{4\mu^2}\pi_{44}e_3^2 + \frac{1}{4\mu^2}\pi_{33}e_4^2 \\ &\quad + \frac{1}{2\mu}\pi_{34}(e_1^2 + e_2^2) \\ &= \frac{1}{2}(e_1^2 - e_2^2)(\pi_{11} - \pi_{22}) + \frac{1}{2\mu}\pi_{34}(e_1^2 + e_2^2) + 2\pi_{12}e_1e_2 \\ &\quad - \frac{1}{\mu}[\pi_{14}e_1e_3 + \pi_{13}e_1e_4 + \pi_{24}e_2e_3 + \pi_{23}e_2e_4] \\ &\quad + \frac{1}{4\mu^2}[\pi_{44}e_3^2 + \pi_{33}e_4^2 + 2\mu(\pi_{11} + \pi_{22})e_3e_4]. \end{aligned}$$

An energy inequality will be obtained by computing $\int_{\mathcal{D}}(\Box\phi)(X\phi)dV$ for some domain $\mathcal{D}$, using the key formula and the Stokes formula: This yields boundary terms and interior terms, which we now discuss separately.

4.4 Energy

Suppose we compute the integral $\int_{\mathcal{D}}(\Box\phi)(X\phi)dV$ for some domain $\mathcal{D}$, using the Stokes formula. The boundary terms that we obtain are

$$\int_{\partial\mathcal{D}} Q(N, X)dv,$$

where N is the unit outgoing normal to $\partial\mathcal{D}$, and dv its volume element.

1. The most common case is, of course, that of a split metric $g = -dt^2 + g_{ij}dx^i dx^j$, when $\mathcal{D} = \{(x, t), 0 \leq t \leq T\}$ is a strip bounded by the hypersurfaces Σ_0 and Σ_T, and $X = -\partial_t$ (we put a minus sign because, with the normalization of g, the wave equation is minus the usual one). The boundary integral is then equal to $E(T) - E(0)$, with

$$E(T) = E_\phi(T) = \int_{\Sigma_T} Q(\partial_t, \partial_t)dv = \tfrac{1}{2}\int_{\Sigma_T}[(\partial_t\phi)^2 + g^{ij}\partial_i\phi\partial_j\phi]dv.$$

This defines the "**energy of ϕ at time T**", which controls (with some weight due to the coefficients g^{ij} and to dv) the L^2 norm of $\nabla\phi$ at time T.

2. Suppose now that the domain $\mathcal{D}$ is bounded by a portion Σ'_T of Σ_T, a portion Σ'_0 of Σ_0 and some "lateral boundary" Λ; in this case, the boundary terms are

$$E(T) - E(0) - \int_\Lambda Q(N, \partial_t)dv,$$

the last term being always nonnegative as long as the lateral boundary of $\mathcal{D}$ has a timelike past oriented normal N. We recognize here the condition for $\mathcal{D}$ to be a domain of determination for $\Box$ (see [9] for instance).

Usually, this last term is neglected in proving energy inequalities, but this is not necessarily a good idea. Suppose, for instance, that, for a given general metric g, we work with a null frame associated with two optical functions u and $\underline{u}$ as in section 2.3.3. Let $\mathcal{D}$ be the domain enclosed between two spacelike hypersurfaces Σ_0 and Σ_T (playing here the role of horizontal planes) and the surface of an incoming light cone $\{\underline{u} = \underline{u}_0\}$ (with $\underline{u}_0 > 0$). On the lateral boundary Λ of $\mathcal{D}$, an exterior unit normal is $N = \alpha^{-1}\nabla\underline{u}$ (with $\alpha = ||\nabla\underline{u}||$); if we take $X = \frac{1}{2}(\nabla u + \nabla\underline{u})$ (which looks like $-\partial_t$), the energy density integrated on Λ is

$$\begin{aligned}(2\alpha)^{-1}Q(\underline{L}, L+\underline{L}) &= (8\alpha\Omega^2)^{-1}Q(e_3, e_3+e_4)\\ &= (8\alpha\Omega^2)^{-1}[e_1(\phi)^2+e_2(\phi)^2+e_3(\phi)^2].\end{aligned}$$

In other words, the energy yields a (weighted) L^2 control of all derivatives of ϕ which are *tangential* to Λ. In the approach of [29], for instance, the authors do not introduce any t-coordinates, and never consider the "energy at time t"; they integrate on domains bounded by incoming cones, and use the energy on these cones as we just explained.

3. More generally, as long as X is timelike past oriented and $\mathcal{D}$ is a domain of determination of $\Box$ (that is, the exterior normal N to the upper part of the boundary of $\mathcal{D}$ is also timelike past oriented), the boundary integral on the upper part of $\partial\mathcal{D}$ has a nonnegative energy density $Q(N, X)$. For the Minkowski metric, for instance, the choice of the timelike past oriented $X = -K_0 = -(r^2+t^2)\partial_t - 2rt\partial_r$ leads to the well-known "conformal inequality" (to which we will return later).

4.5 Interior terms and positive fields

In the previous section, we discussed only the sign of the boundary terms arising from the computation of $\int_{\mathcal{D}}(\Box\phi)(X\phi)dV$ for some domain $\mathcal{D}$, and the concept of (positive) energy. In general, however, we also have to deal with the interior terms

$$\int_{\mathcal{D}} Q_{\alpha\beta}{}^{(X)}\pi^{\alpha\beta}dV \equiv \int_{\mathcal{D}} Q\pi dV.$$

There are basically two different strategies for dealing with these terms:

(i) control by brute force, and
(ii) discussion of the signs of the terms.

1. Two examples of brute force control

a. ***Using the Gronwall lemma*** The idea here is to bound the interior integrand $Q\pi$ by the integrand of the energy; in the simple case of a split metric and a domain $\mathcal{D} = \{0 \le t \le T\}$, for instance, suppose that we can obtain the bound

$$\int_{\Sigma_t} |Q^{(\partial_t)}\pi|dv \le 2C(t)E(t).$$

We will write our inequality

$$E(T) - E(0) \leq \int (\Box\phi)(X\phi)dV + \int_0^T C(t)E(t)dt,$$

and the Gronwall lemma yields

$$E(T) \leq \{E(0) + \int_{\mathcal{D}} |(\Box\phi)(X\phi)|dV\} \exp\left(\int_0^T C(t)dt\right).$$

The interior terms have thus disappeared from the energy inequality, but at the cost of the amplification factor $\exp(\int_0^T C(t)dt)$. When dealing with problems that are global in time, for example, this can be disastrous if C is not integrable.

b. *Weighted inequality* An approach essentially equivalent to that of **a**. is as follows. We replace X by fX for a well-chosen function f. Using the formula of section 3.2, we have then

$$Q^{(fX)}\pi = fQ\pi + 2Q(\nabla f, X).$$

A common choice is to take $f = e^{\lambda a}$ for some function a and some real number λ:

$$Q^{(fX)}\pi = e^{\lambda a}[Q\pi + 2\lambda Q(\nabla a, X)].$$

If X and ∇a are both timelike past oriented (for instance, $a = t$), reasonable assumptions make it possible to obtain

$$Q\pi + 2\lambda Q(\nabla a, X) \geq 0$$

for large enough λ. In this case, the interior terms have the good sign and we have a weighted inequality. The drawback of this well-known approach is, of course, that one has to keep the weight $e^{\lambda a}$ in the formula for the energy.

2. Sign control of interior terms In applications to global problems, the brute force strategy is generally too crude, so we discuss now the sign strategy. If the multiplier X happens to be a Killing field, ${}^{(X)}\pi \equiv 0$ and the interior term is identically 0. For instance, this is the case for the multiplier $X = \partial_t$ for the flat metric.

Leaving aside this trivial and miraculous case, we prove now the following theorem.

Theorem *For any C^2 function R, the following identity holds*

$$\int_{\mathcal{D}} R|\nabla\phi|^2 dV = \frac{1}{2}\int_{\mathcal{D}} \phi^2(\Box R)dV - \int_{\mathcal{D}} R\phi(\Box\phi)dV$$
$$- \frac{1}{2}\int_{\partial\mathcal{D}} \phi^2\langle N, \nabla R\rangle dv + \int_{\partial\mathcal{D}} R\phi\langle N, \nabla\phi\rangle dv.$$

Proof Observe that, for any two functions f and h,

$$\operatorname{div} f\nabla h = D_\alpha(f\nabla h)^\alpha = \partial_\alpha f \partial^\alpha h + f\Box h.$$

Using this formula with $f = R\phi$ and $h = \phi$, or $f = \phi^2$ and $h = R$, and integrating with the help of the Stokes formula, we get the result. ◇

We give a first example of how this formula can be used.

Example 1 Suppose the multiplier X is conformal Killing; then

$$Q_{\alpha\beta}\pi^{\alpha\beta} = \lambda Q_{\alpha\beta}g^{\alpha\beta} = \lambda tr\, Q = -\lambda|\nabla\phi|^2.$$

One should not be misled by the expression $|\nabla\phi|^2$: it is by no means a positive quantity! Using the above theorem, one can eliminate these bad terms. However, this transformation has two drawbacks:

(i) it produces new interior term $\int_{\mathcal{D}} \phi^2(\Box\lambda)dV$, which must be nonnegative if we are to obtain an energy inequality;
(ii) it produces additionnal boundary terms involving ϕ and $\nabla\phi$, which can spoil the positivity of the energy.

A typical example is the choice $-X = K_0$ of a multiplier for the flat metric in a domain $\mathcal{D} = \{0 \le t \le T\}$. We have the identity

$$\begin{aligned} 2(\Box\phi)(X\phi) &= \partial_t[(r^2+t^2)((\partial_t\phi)^2 + \sum(\partial_i\phi)^2) + 4rt(\partial_t\phi)(\partial_r\phi)] \\ &\quad + \sum \partial_i[\cdots] + 4t[(\partial_t\phi)^2 - \sum(\partial_i\phi)^2]. \end{aligned}$$

Transforming the interior term $-4t|\nabla\phi|^2$ yields the identity

$$\int_{\mathcal{D}} (\Box\phi)(X\phi - 2t\phi)dxdt = \tilde{E}(T) - \tilde{E}(0),$$

where the modified energy $\tilde{E}$ is now

$$\begin{aligned} \tilde{E}(T) = \tfrac{1}{2}\int_{\Sigma_T} &\{(r^2+t^2)[(\partial_t\phi)^2 + \sum(\partial_i\phi)^2] + 4rt(\partial_t\phi)(\partial_r\phi) \\ &+ 4rt\phi(\partial_t\phi) - 2\phi^2\}dx. \end{aligned}$$

In this example, we do not have to worry about the term $\int \phi^2(\Box\lambda)dV$, since $\lambda = -4t$ and $\Box\lambda = 0$. It is a delicate task to prove that the modified energy $\tilde{E}$ is indeed positive (see [9] or [24] for instance). One finally obtains, for some constant $C > 0$,

$$\begin{aligned} C^{-1}\int [(S\phi)^2 + |H\phi|^2 + |R\phi|^2 + \phi^2](x,t)dx \le \tilde{E}(t) \le C\int [(S\phi)^2 \\ + |H\phi|^2 + |R\phi|^2 + \phi^2](x,t)dx. \end{aligned}$$

The formula on $R|\nabla\phi|^2$ can also be used to eliminate *some part* of the interior terms, as shown in the following example.

Example 2 It can happen that the interior term $Q\pi$ is nonnegative, *up to a multiple of* $|\nabla\phi|^2$. A typical example is that of the Morawetz inequality for the flat metric (see, for instance, [9]): taking $X = -\partial_r$, we have the identity

$$2(\Box\phi)(X\phi) = \partial_t[\cdots] + \sum \partial_i[\cdots] + \frac{2}{r}\left[\sum(\partial_i\phi)^2 - (\partial_r\phi)^2\right] - \frac{2}{r}|\nabla\phi|^2.$$

Of course, in this case the corresponding energy will not be positive, since X is spacelike. The boundary terms will have to be controlled separately, using the standard energy inequality (corresponding to $X = -\partial_t$). Note that in this example $\lambda = 2/r$, and $\Box(1/r)$, which is zero for $r > 0$, is singular at the origin. As a result, the new interior term $\int_{\mathcal{D}} \phi^2(\Box\lambda)dV$ is $\int_0^T \phi^2(0,t)dt$.

The preceding examples suggest the following definition.

Definition A **positive field** X for the metric g is a field such that, for some R,

$$I = Q_{\alpha\beta}{}^{(X)}\pi^{\alpha\beta} + R|\nabla\phi|^2$$

is a positive quadratic form in $\nabla\phi$.

Lemma *The fact that X is positive depends only on X and on the conformal class of the metric g.*

Proof Set $\tilde{g} = e^\lambda g$, that is, $\tilde{g}_{\alpha\beta} = e^\lambda g_{\alpha\beta}$. Then

$$\tilde{g}^{\alpha\beta} = e^{-\lambda}g^{\alpha\beta},\ \tilde{g}(\tilde{\nabla}\psi, \tilde{\nabla}\psi) = e^{-\lambda}|\nabla\psi|^2,$$

and consequently $\tilde{Q}_{\alpha\beta} = Q_{\alpha\beta}$. Now ${}^{(X)}\tilde{\pi} = \mathcal{L}_X\tilde{g} = e^\lambda[\mathcal{L}_X g + (X\lambda)g]$, hence finally

$$\begin{aligned}\tilde{Q}^{(X)}\tilde{\pi} + \tilde{R}\tilde{g}(\tilde{\nabla}\psi, \tilde{\nabla}\psi) &= e^\lambda[Q^{(X)}\pi + (X\lambda)\mathrm{tr}\, Q + \tilde{R}e^{-2\lambda}|\nabla\psi|^2] \\ &= e^\lambda[Q^{(X)}\pi + R|\nabla\psi|^2 + |\nabla\psi|^2(-R + \tilde{R}e^{-2\lambda} - X\lambda)].\end{aligned}$$

It is enough to choose $\tilde{R} = e^{2\lambda}(X\lambda + R)$ to show that X is also positive for $\tilde{g}$. ◇

If X is a positive field and we perform the transformation of the above theorem on the term $R|\nabla\phi|^2$, we obtain, with $I = Q\pi + R|\nabla\phi|^2 \geq 0$,

$$\int_{\mathcal{D}} Q^{(X)}\pi dV = \int_{\mathcal{D}} (I - R|\nabla\phi|^2)dV = \int_{\mathcal{D}} (I - \tfrac{1}{2}\phi^2\Box R)dV + \int_{\partial\mathcal{D}} \cdots dv.$$

If $\Box R \leq 0$, the integrand in $\mathcal{D}$ is positive and the boundary terms are well behaved, we may hope in the end to obtain an energy inequality. If $\Box R$ has no special sign, we cannot say anything about the integrand in $\mathcal{D}$, but it can happen that

$$\int_{\mathcal{D}} I dV \geq \tfrac{1}{2} \int_{\mathcal{D}} \phi^2 \Box R dV.$$

An inequality of this type is called a Poincaré inequality. A typical example occurs when trying to prove a Morawetz type inequality for the Schwarzschild metric.

Example 3 Let us first introduce, in the exterior (of the black hole) region $r > 2m$, the coordinate

$$r^* = r + 2m \log(r - 2m) - 3m - 2m \log m,$$

where the funny normalization is meant to get $r^* = 0$ for $r = 3m$ (this sphere is called the photon sphere, see [20], [23]). With the coordinates (t, r^*, θ, ϕ), the wave equation is

$$\Box\phi = \left(1 - \frac{2m}{r}\right)^{-1} [-\partial_t^2\phi + r^{-2}\partial_{r^*}(r^2\partial_{r^*}\phi)] + r^{-2}\Delta_{S^2}\phi.$$

Consider now a domain $\mathcal{D} = \{0 \leq t \leq T\}$, and choose the multiplier $X = f(r^*)\partial_{r^*}$. A staightforward computation (see, for instance, [19]) gives

$$\int_{\mathcal{D}} Q\pi dV = \int_0^T \int_{-\infty}^{+\infty} \int_{S^2} \left[\frac{f'}{1-\nu}(\partial_{r^*}\phi)^2 + \frac{r-3m}{r^2} f |\nabla\!\!\!/ \phi|^2 \right.$$
$$\left. - \frac{1}{2}\left(f' + 2f\frac{1-\nu}{r}\right)|\nabla\phi|^2 \right] r^2(1-\nu) dt dr^* d\sigma_{S^2}.$$

Here, $\nu = 2m/r$ and $\nabla\!\!\!/ \phi$ is the gradient of the restriction of ϕ to the spheres

$$|\nabla\!\!\!/ \phi|^2 = e_1(\phi)^2 + e_2(\phi)^2.$$

We see that if f is increasing and vanishes at $r^* = 0$, the field $f\partial_{r^*}$ is positive. Eliminating the $|\nabla\phi|^2$ term as above, we are left with the expression

$$I = \int_0^T \int_{-\infty}^{+\infty} \int_{S^2} \left[\frac{f'}{1-\nu}(\partial_{r^*}\phi)^2 + \frac{r-3m}{r^2} f |\nabla\!\!\!/ \phi|^2 \right.$$
$$\left. - \frac{1}{4}\left(\Box\left(f' + 2f\frac{1-\nu}{r}\right)\right)\phi^2 \right] r^2(1-\nu) dt dr^* d\sigma_{S^2}.$$

Since

$$\Box\left(f' + 2f\frac{1-\nu}{r}\right) = \frac{1}{1-\nu} f''' + \frac{4}{r} f'' - 8mr^{-2}(r-2m)^{-1} f'$$
$$- 2mr^{-3}(3-\nu) f,$$

we easily see that f cannot be chosen to ensure the negativity of this coefficient.

The only thing we can do is to try to use the strength of the first term $(f'/(1-\nu))(\partial_{r^*}\phi)^2$ to compensate for the bad sign of the ϕ^2 term. The Poincaré inequality we use is the consequence of the following trivial identity, valid for any C^1 function α,

$$\int f'(\partial_{r^*}\phi)^2 r^2 dr^* = \int f'(\partial_{r^*}\phi + \alpha\phi)^2 r^2 dr^* + \int \phi^2 \left[f''\alpha + f'\left(\alpha' - \alpha^2 + 2\alpha\frac{1-\nu}{r}\right)\right] r^2 dr^*.$$

Dropping the first term of the right-hand side, we obtain a Poincaré inequality, depending on some unknown function α still to be chosen. It turns out that we also need to use the contribution from the $|\not\nabla\phi|^2$ term. For this, we decompose ϕ into spherical harmonics ϕ_l. Finally, one can prove that there is some l_0, some function f, and some function α such that, if $\phi_l = 0$ for $l \leq l_0$,

$$I \geq \int \langle r^* \rangle^{-3-0} \phi^2 dV.$$

One can consult [19] or [20] for details.

3. Interior terms and Poisson bracket Note that

$$(\Box\phi)(X\phi) = \operatorname{div}(X\phi)\nabla\phi - \operatorname{div}\phi\nabla(X\phi) + \phi[\Box, X]\phi + \phi X\Box\phi.$$

Integrating this identity in some domain $\mathcal{D}$, we see that the interior term $\int_{\mathcal{D}} Q\pi\, dV$ must correspond, modulo $|\nabla\phi|^2$, to the quadratic form $\int_{\mathcal{D}}([\Box, X]\phi)\, \phi dV$. In fact, we have

$$Q\pi = \partial_\alpha\phi\partial_\beta\phi\pi^{\alpha\beta} - \tfrac{1}{2}|\nabla\phi|^2 \operatorname{tr}\pi.$$

Using the expression for $\pi^{\alpha\beta}$ given in section 4.2,

$$(\partial_\alpha\phi)(\partial_\beta\phi)\pi^{\alpha\beta} = (\partial_\alpha\phi)(\partial_\beta\phi)[2\partial^\alpha(X^\beta) - X(g^{\alpha\beta})].$$

This quadratic form corresponds to the operator with symbol $q = \xi_\alpha\xi_\beta[2\partial^\alpha (X^\beta) - X(g^{\alpha\beta})]$. We note that q is precisely the **Poisson bracket**

$$q = \{g^{\alpha\beta}\xi_\alpha\xi_\beta, X^\gamma\xi_\gamma\}$$

of the symbol of $\Box$ with the symbol of X. Thus the formula for $\pi^{\alpha\beta}$ of section 4.2 provides a connexion between deformation tensors and Poisson brackets.

4.6 Maxwell equations

In the Maxwell equations

$$dF = 0,\ D^\alpha F_{\alpha\beta} = 0,$$

the unknown object is a 2-form F (that is, an antisymmetric 2-tensor). For the definition of the exterior derivative d, see Spivak [46]. In the flat case, to connect this formulation with the usual formulation, set

$$E_i = -F_{0i},\ i = 1, 2, 3,\ H^1 = -F_{23},\ H^2 = -F_{31},\ H^3 = -F_{12},$$

thus defining the "**electric field**" E and the "**magnetic field**" H. Then

$$dF = 0 \Leftrightarrow \operatorname{div} H = 0,\ \partial_t H - \operatorname{curl} E = 0,$$
$$D^\alpha F_{\alpha\beta} = 0 \Leftrightarrow \operatorname{div} E = 0,\ \partial_t E + \operatorname{curl} H = 0.$$

4.6.1 Duality

A convenient way of doing many computations is to introduce the **dual form**

$$*F_{\mu\nu} = \tfrac{1}{2}\epsilon_{\mu\nu\alpha\beta}F^{\alpha\beta},$$

where ϵ is the volume form (see section 2.1). This summation looks complicated, but it is enough to note that for given (μ, ν), there are only two indices, say $(\underline{\mu}, \underline{\nu})$, different from μ and from ν, so that

$$*F_{\mu\nu} = \epsilon_{\mu\nu\underline{\mu}\underline{\nu}}F^{\underline{\mu}\underline{\nu}},$$

the right-hand side being in this case only one term (no sum!). Also,

$$**F_{\mu\nu} = \tfrac{1}{4}\epsilon_{\mu\nu\alpha\beta}\epsilon^{\alpha\beta\gamma\delta}F_{\gamma\delta} = |\epsilon|F_{\mu\nu} = -F_{\mu\nu}.$$

In the flat case, for instance,

$$*F_{01} = \epsilon_{0123}F^{23} = F_{23} = -H^1,$$

which shows that the electric field of $*F$ is just the magnetic field of F; also, in view of the relation above, the magnetic field of $*F$ is minus the electric field of F.

Lemma *For any vector field X,*

$$D_X * F = *D_X F.$$

Proof This may seem obvious, since $*$ is defined using the metric only, and $D_X g = 0$; we give, however, a self-contained proof. First, we observe $D_X\epsilon = 0$:

if e_α is an orthonormal basis,

$$X(\epsilon(e_1, e_2, e_3, e_4)) = 0 = (D_X\epsilon)(e_1, e_2, e_3, e_4) + \epsilon(D_X e_1, e_2, e_3, e_4) + \cdots,$$

and the last three terms are zero since $D_X e_\alpha$ has no component on e_α. Next, using the definitions, we get

$$\begin{aligned} X(*F_{\mu\nu}) &= (D_X * F)_{\mu\nu} + (*F)(D_X\partial_\mu, \partial_\nu) + (*F)(\partial_\mu, D_X\partial_\nu) \\ &= [\epsilon(D_X\partial_\mu, \partial_\nu, \partial_\alpha, \partial_\beta) + \epsilon(\partial_\mu, D_X\partial_\nu, \partial_\alpha, \partial_\beta) + \epsilon(\partial_\mu, \partial_\nu, D_X\partial_\alpha, \partial_\beta) \\ &\quad + \epsilon(\partial_\mu, \partial_\nu, \partial_\alpha, D_X\partial_\beta)]F^{\alpha\beta} + \epsilon_{\mu\nu\alpha\beta}[(D_X F)^{\alpha\beta} \\ &\quad + F(D_X\partial^\alpha, \partial^\beta) + F(\partial^\alpha, D_X\partial^\beta)]. \end{aligned}$$

Cancelling the terms and using the symmetries, we finally obtain

$$\begin{aligned} [(D_X * F)_{\mu\nu} - (*D_X F)_{\mu\nu}] &= 2\epsilon(\partial_\mu, \partial_\nu, D_X\partial_\alpha, \partial_\beta)F^{\alpha\beta} + 2\epsilon_{\mu\nu\alpha\beta}F(D_X\partial^\alpha, \partial^\beta) \\ &= I + II. \end{aligned}$$

Taking $X = \partial_\gamma$ and using the formula for the Christoffel symbols, we see that the first term I is just

$$I = \epsilon_{\mu\nu\lambda\beta}g^{\lambda\lambda'}(-\partial_{\lambda'}g_{\alpha\gamma} + \partial_\alpha g_{\gamma\lambda'} + \partial_\gamma g_{\alpha\lambda'})F^{\alpha\beta}.$$

On the other hand,

$$D_\gamma\partial^\alpha = D_\gamma(g^{\alpha\beta}\partial_\beta) = \partial_\gamma g^{\alpha\beta}\partial_\beta + g^{\alpha\beta}\Gamma_{\lambda\gamma\beta}\partial^\lambda.$$

Hence, using $\partial_\gamma g^{\alpha\beta} = -g^{\alpha\alpha'}g^{\beta\beta'}\partial_\gamma g_{\alpha'\beta'}$,

$$2(D_\gamma\partial^\alpha)_\lambda = g^{\alpha\beta}(-\partial_\gamma g_{\lambda\beta} + \partial_\beta g_{\gamma\lambda} - \partial_\lambda g_{\gamma\beta}).$$

This gives us the expression for the second term II

$$II = \epsilon_{\mu\nu\alpha\beta}g^{\alpha\delta}(\partial_\delta g_{\gamma\lambda} - \partial_\lambda g_{\gamma\delta} - \partial_\gamma g_{\lambda\delta})F^{\lambda\beta}.$$

We see, changing the indices, that each term in I is the opposite of the corresponding term in II. ◇

The duality makes it possible to rewrite Maxwell equations in a symmetric way.

Theorem *The Maxwell equations can be written*

$$dF = 0, \; d * F = 0.$$

Proof Let us admit the following formula for the exterior derivative of a 2-form (see [23] for details)

$$F = F_{\alpha\beta}dx^\alpha \wedge dx^\beta, \; dF = D_\gamma F_{\alpha\beta}dx^\gamma \wedge dx^\alpha \wedge dx^\beta.$$

Using this formula for $*F$ and the commutation formula above, we obtain

$$\begin{aligned} d(*F) &= (*D_\gamma F)_{\alpha\beta} dx^\gamma \wedge dx^\alpha \wedge dx^\beta \\ &= \epsilon_{\alpha\beta\mu\nu} D_\gamma F^{\mu\nu} dx^\gamma \wedge dx^\alpha \wedge dx^\beta. \end{aligned}$$

Now fix the index δ and look at the sum S of the terms involving (in some order) the three remaining indices $a < b < c$. We can write

$$\begin{aligned} \frac{S}{2} &= \epsilon_{ab\mu\nu} D_c F^{\mu\nu} dx^c \wedge dx^a \wedge dx^b \\ &\quad + \epsilon_{bc\mu\nu} D_a F^{\mu\nu} dx^a \wedge dx^b \wedge dx^c + \epsilon_{ac\mu\nu} D_b F^{\mu\nu} dx^b \wedge dx^a \wedge dx^c. \end{aligned}$$

Consider the first term of the right-hand side. Either μ or ν has to be c, and then the other index is necessarily δ; hence this first term is $2\epsilon_{abc\delta} D_c F^{c\delta} dx^c \wedge dx^a \wedge d^b$. Handling the two other terms similarly, we get

$$\frac{S}{4} = \epsilon_{abc\delta} dx^a \wedge dx^b \wedge dx^c [D_\alpha F^{\alpha\delta}],$$

which finishes the proof. ◇

4.6.2 Energy formalism

For Maxwell equations, we can also define an energy–momentum tensor

$$Q_{\alpha\beta} = 2F_{\alpha\gamma} F_\beta^\gamma - \tfrac{1}{2} g_{\alpha\beta} |F|^2, \; |F|^2 = F_{\lambda\mu} F^{\lambda\mu}.$$

If F is a solution of the Maxwell equations, one can prove, in a similar way to that used in section 4.1,

$$\forall \beta, \; D^\alpha Q_{\alpha\beta} = 0.$$

Also, Q has the same positivity property as in the case of the wave equation, since, in any null frame,

$$Q_{33} \geq 0, \; Q_{44} \geq 0, \; Q_{34} \geq 0.$$

These inequalities are proved in section 5.3. The only difference with the case of the wave equation is that

$$\operatorname{tr} Q = Q_\alpha^\alpha = 2F_{\alpha\gamma} F^{\alpha\gamma} - \tfrac{1}{2} \times 4 \times |F|^2 = 0.$$

The method for proving energy inequalities is exactly the same as before. We choose a timelike multiplier X, set $P_\alpha = Q_{\alpha\beta} X^\beta$. If F is a solution of Maxwell equations, we have

$$D^\alpha P_\alpha = \tfrac{1}{2} Q_{\alpha\beta} \pi^{\alpha\beta}.$$

Integrating in some domain $\mathcal{D}$ yields the identity

$$\tfrac{1}{2}\int_{\mathcal{D}} Q\pi dV = \int_{\partial\mathcal{D}} Q(N, X)dv.$$

All we have said about energy and interior terms extends also to this case, with the improvement that if X is conformal Killing, the interior terms are identically zero.

In chapter 8, we will discuss briefly the analogous and more complicated case of Bianchi equations, for which an energy formalism also exists.

5

The good components

5.1 The problem

Let g be a given Lorentzian metric on $\mathbf{R}^4$, say, not too far from the flat Minkowski metric. Consider a solution ϕ of the wave equation $\Box_g \phi = 0$ with smooth Cauchy data on $\{t = 0\}$ rapidly decaying at infinity. By analogy with the flat case (see the introduction), we suspect that some derivatives of ϕ behave better than others, meaning here that they have better decay properties at infinity. These "good derivatives" of ϕ (again, "good" is meant here in the sense of having a better decay at infinity) are some of the components of $\nabla\phi$ in an appropriate null frame; in the flat case, these good derivatives are $e_1(\phi)$, $e_2(\phi)$, $e_4(\phi)$. For a general metric g, the question we ask is the following:

> How is a null frame (e_α) that would "capture" the good components of $\nabla\phi$ to be chosen?

The same question arises when dealing with Maxwell equations:

> What frame is going to capture the good components of the 2-form F?

There are, as far as we know, three approaches to this problem:

(1) weighted "standard" energy inequalities,
(2) conformal energy inequalities,
(3) commutation with modified Lorentz fields.

In the first approach, one establishes an improved version of the "standard" energy inequality (by this we mean the inequality corresponding to the multiplier ∂_t in the flat case); such an inequality yields, besides the usual bound of the energy at time t, a bound of the (weighted) L^2 norm in both variables x and t of some special derivatives of ϕ. In other words, the L^2 norm in x of these special derivatives is not just bounded in t, it is an L^2 function of t also.

This fact identifies these special derivatives as the "good derivatives" of our problem.

In the second approach, one generalizes the conformal inequality of the flat case. Recall that, in the flat case, the conformal inequality gives a bound of the modified energy $\tilde{E}$, which is equivalent to the sum of the L^2 norms of the special fields S, R_i, H_i,

$$\tilde{E}(t) \sim \int_{\Sigma_t} [(S\phi)^2 + |R\phi|^2 + |H\phi|^2 + \phi^2](x,t)dx.$$

Through the identities

$$L = (r+t)^{-1}[S + \sum \omega_i H_i],\ \omega \wedge \partial = \frac{R}{r} = t^{-1}\omega \wedge H,$$

one obtains a t^{-1} decay of the L^2 norms in x of the special derivatives $L\phi, (R/r)\phi$ of ϕ, a fact which identifies these derivatives as the "good derivatives." In the general case, there are generalizations of the conformal inequality: the structure of the corresponding modified energy $\tilde{E}$ will display the good derivatives.

Finally, the third approach is to commute the Lorentz vector fields $Z = \partial_\alpha, S, R_i, H_i$ with $\square$, and then to use the standard energy inequality. In this way one obtains a bound of the L^2 norms in x of $\nabla Z\phi$ (or, equivalently, of $Z\nabla\phi$), and one proceeds as above to identify the good derivatives from the Z fields. In the general case, one constructs modified Lorentz fields $\tilde{Z}$, for which the commutators $[\square, \tilde{Z}]$ are small, and obtains the good derivatives from the fields $\tilde{Z}$.

We give some precise statements for the first two cases, postponing the discussion of commutators to the next chapter.

5.2 An important remark

In this chapter, we put emphasis on the decay at infinity of global solutions. However, as we shall see in the last chapter, the use of appropriate null frames is not limited to problems involving global solutions and their decay at infinity. It shoud be considered as a universal method for all hyperbolic problems involving only one Lorentzian metric (such as Einstein equations); this is beautifully explained by Christodoulou in the prologue of [16]. If decay is not the problem, what are the guidelines to choose a good null frame? The basic principle of the construction of null hypersurfaces and optical functions was sketched in section 3.3. We will see in the last chapter concrete examples of such constructions; the exact choices and the benefit thereof depend on the context.

5.3 Ghost weights and improved standard energy inequalities

1. The key idea We explain here the key idea in obtaining a "good" energy inequality. Let g be given as usual, and suppose we have chosen a null frame which is *suspected* to capture the good components of $\nabla\phi$, in the sense that $e_1(\phi)$, $e_2(\phi)$, $e_4(\phi)$, the "good" derivatives of ϕ, decay better than $e_3(\phi)$. For a given timelike multiplier X, we have to handle the interior terms $\int_{\mathcal{D}} Q^{(X)}\pi \, dV$, as explained in section 4.3 Writing $Q\pi$ in our null frame, we see that all terms, except $\pi_{44}e_3(\phi)^2$, involve at least one "good" derivative. Since we want $Q\pi$ to be as small as possible, we choose X such that ${}^{(X)}\pi_{44} = 0$.

Key idea Choose the multiplier X such that ${}^{(X)}\pi_{44} = 0$.

However, this does not tell us how to guess the good null frame! What we do is to reversed the problem: according to our geometric intuition, we choose a reasonable null frame, and then give conditions on g which ensure that this null frame is indeed a good null frame. Though this sketch may sound a little strange, we will see in examples how this method works.

We start with two cases, giving complete proofs of the corresponding theorems.

2. The wave equation in the quasiradial case We make the assumptions described in chapter 2 for the quasiradial case

$$g^{00} = -1, \; g^{0i}\omega_i = 0, \; \omega_i = \omega^i.$$

We take our null frame to be

$$(e_1, e_2, e_3 = T - N, e_4 = T + N),$$

where (e_1, e_2) form an orthonormal basis (for g) on the standard spheres of $\mathbf{R}^3$ (for constant t), and

$$T = -\nabla t = \partial_t - g^{0i}\partial_i, \; N = \nabla r / ||\nabla r||.$$

We set $c = ||\nabla r|| = (g^{ij}\omega_i\omega_j)^{1/2}$, and define the second fundamental form k of Σ_t by

$$k(X, Y) = -\langle D_X T, Y\rangle.$$

Recall the formula for the components of k,

$$k_{ij} = -\tfrac{1}{2} g^{0\alpha}(\partial_i g_{\alpha j} + \partial_j g_{\alpha i} - \partial_\alpha g_{ij}).$$

Finally, we define the energy at time t to be

$$E(t) = \tfrac{1}{2} \int_{\Sigma_t} [(T\phi)^2 + (N\phi)^2 + |\not\nabla\phi|^2] dv,$$

recalling the notation $|\not\nabla\phi|^2 = e_1(\phi)^2 + e_2(\phi)^2$.

Theorem *Assume that the components of k satisfy, for some $\epsilon > 0$,*

(i) $\langle t - r\rangle^{1+\epsilon}[k_{1N}^2 + k_{2N}^2 + (k_{11} + k_{22})^2] \in L_t^1 L_x^\infty$,
(ii) $\langle t - r\rangle^{1+\epsilon}[|Tc/c| + |k_{1N}| + |k_{2N}| + |k_{11}| + |k_{12}| + |k_{22}|] \in L_{x,t}^\infty$.

Then, for some constant $C = C_\epsilon$ and all $T \geq 0$,

$$E(T) + \int_{0\leq t\leq T} \langle t - r\rangle^{-1-\epsilon}[e_4(\phi)^2 + |\not\nabla\phi|^2] dV$$
$$\leq CE(0) + C\int_{0\leq t\leq T} |\Box\phi||T\phi| dV + C\int_0^T A(t)E(t)dt.$$

The amplification factor A is

$$A(t) = \left\|\frac{Tc}{c}\right\|_{L_x^\infty} + ||\langle t - r\rangle^{-1}(c - 1)||_{L_x^\infty}.$$

In particular, if $A \in L_t^1$, $e_1(\phi)$, $e_2(\phi)$, $e_4(\phi)$ are the good components of $\nabla\phi$.

Note that assumptions (i) and (ii) do not require the derivatives of the components of the metric to be in $L_t^1 L_x^\infty$.

Proof The proof of the theorem is an application of the key idea above, combined with the use of the weight e^a, $a = a(t - r)$.

(a) We choose the multiplier $X = e^a T$, and set $\pi = {}^{(T)}\pi$: from the formula of section 4.2,

$$Q^{(X)}\pi = e^a[Q\pi + 2Q(\nabla a, T)].$$

Since

$$\nabla a = a'\nabla(t - r) = -a'(T + cN) = -a'(T + N + (c - 1)N),$$

and $2Q(T, T) = (T\phi)^2 + (N\phi)^2 + |\not\nabla\phi|^2$, we can compute the additional terms due to the weight

$$Q(T + N, T) = \tfrac{1}{2}(e_4(\phi)^2 + |\not\nabla\phi|^2),$$
$$-Q^X\pi = e^a[-Q\pi + a'(e_4(\phi)^2 + |\not\nabla\phi|^2) + 2a'(c - 1)(T\phi)(N\phi)].$$

The idea of the "ghost weight" is to choose $a'(s) = A\langle s\rangle^{-1-\epsilon}$. In this way, a is bounded and the weight e^a *disappears* from the inequality (leaving behind only constants depending on ϵ and A); on the other hand, choosing A big enough will give us plenty of good derivatives.

(b) We now separate all terms of $Q\pi$ into three categories:

(i) the terms containing two good derivatives, which have the coefficients π_{34}, π_{33}, π_{a3}, π_{ab} $(1 \le a, b \le 2)$,

(ii) the terms containing only one good derivative, which have the coefficients π_{a4}, $\pi_{11} + \pi_{22}$,

(iii) the bad term $\pi_{44} e_3(\phi)^2$.

To handle the terms in the first two categories, it is enough to assume respectively

$$\langle t - r\rangle^{1+\epsilon}[|\pi_{34}| + |\pi_{33}| + \sum |\pi_{a3}| + \sum |\pi_{ab}|] \in L^\infty_{x,t},$$
$$\langle t - r\rangle^{1+\epsilon}[\sum (\pi_{a4})^2 + (\pi_{11} + \pi_{22})^2] \in L^1_t L^\infty_x.$$

(c) To allow the simple choice of our multiplier, we do not realize exactly $\pi_{44} = 0$ as announced in the "key idea", but $\pi_{44} = -2Tc/c$. In fact, since

$$T = \partial_t - g^{0i}(\partial_i - \omega_i \partial_r) = \partial_t + g^{0i}\left(\omega \wedge \frac{R}{r}\right)_i,$$
$$Nc^{-1} g^{ij}\omega_i \partial_j = c\partial_r - c^{-1} g^{ij}\omega_i \left(\omega \wedge \frac{R}{r}\right)_j,$$

we obtain

$$[T, N] = \frac{Tc}{c} N + \cdots R, \ \langle [T, N], N\rangle = \frac{Tc}{c}.$$

On the other hand, since $D_T T = 0$,

$$\langle [T, N], N\rangle = \langle D_T N, N\rangle - \langle D_N T, N\rangle = k_{NN},$$
$$\pi_{44} = 2\langle D_{T+N} T, T + N\rangle = 2\langle D_N T, T + N\rangle = -2k_{NN}.$$

(d) Finally, it remains to compute the components of π to translate the conditions above on π into the conditions of the theorem on k. We obtain easily

$$\pi_{34} = 2k_{NN} = -\pi_{33}, \pi_{a4} = 2k_{aN} = -\pi_{a3}, \pi_{ab} = 2k_{ab}. \qquad \diamond$$

The same method can be extended to generalize Morawetz type inequalities, see, for instance, [5], [7].

3. The wave equation in the general case We present here a more geometric, but less explicit, result. Assume given an optical function u to which we associate a null frame ($e_1, e_2, e_3 = \underline{L}, e_4 = L$) as explained in chapter 2. Define the corresponding energy at time T to be

$$E(T) = \tfrac{1}{2}\int_{\Sigma_T} \{a[e_4(\phi)]^2 + a^{-1}[e_3(\phi)]^2 + (a + a^{-1})|\nabla\!\!\!/\phi|^2\}dv.$$

Theorem *Let $T = \frac{1}{2}(L + \underline{L})$ and $\pi = {}^{(T)}\pi$. Assume that, for some $\epsilon > 0$, the components of π satisfy the following estimates:*

(i) $\langle u\rangle^{1+\epsilon} a[(\pi_{1L})^2 + (\pi_{2L})^2 + (\pi_{11} + \pi_{22})^2] \in L_t^1 L_x^\infty$,
(ii) $\langle u\rangle^{1+\epsilon}[|\pi_{11}| + |\pi_{22}| + |\pi_{12}| + |\pi_{1\underline{L}}| + |\pi_{2\underline{L}}| + |\underline{\omega}|] \in L_{x,t}^\infty$.

Then, for some $C = C_\epsilon$ and all T,

$$E(T)^{\frac{1}{2}} + \left\{\int_{0\le t\le T} \langle u\rangle^{-1-\epsilon}[(L\phi)^2 + |\nabla\!\!\!/\phi|^2]dtdv\right\}^{\frac{1}{2}}$$
$$\le CE(0)^{\frac{1}{2}} + C\int_0^T ||(a^{\frac{1}{2}} + a^{-\frac{1}{2}})f(\cdot, t)||_{L^2(dv)}(t)dt.$$

The main feature here is that there is *no amplification factor* at all! In comparison with the preceding theorem, we see that the amplification factor A in the preceding theorem came from Tc/c (the error in π_{44}) and $c - 1$ (the error coming from taking $u = t - r$ instead of a *true* optical function).

In particular, in this theorem, the derivatives $e_1(\phi), e_2(\phi), e_4(\phi)$ are identified as the good derivatives.

Proof The components of π can be computed explicitly in terms of the frame coefficients. Following the "key idea," we have arranged for $\pi = {}^{(T)}\pi$

$$\pi_{LL} = \langle D_L(L + \underline{L}), L\rangle = -\langle L + \underline{L}, D_L L\rangle = 0.$$

The other components appearing in the assumptions of the theorem are

$$2\pi_{La} = \langle D_L(L + \underline{L}), e_a\rangle + \langle D_a(L + \underline{L}), L\rangle = 2\underline{\eta}_a - 2\eta_a,$$
$$2\pi_{\underline{L}a} = \langle D_{\underline{L}}(L + \underline{L}), e_a\rangle + \langle D_a(L + \underline{L}), \underline{L}\rangle = 2\eta_a + 2\underline{\xi}_a + 2\eta_a$$
$$= 4\eta_a + 2\underline{\xi}_a,$$
$$\pi_{ab} = \chi_{ab} + \underline{\chi}_{ab}, \; 4\underline{\omega} = \nabla^2 u_{\underline{L}\underline{L}}.$$

The proof of the theorem follows the same lines as before. First, using the multiplier T gives an energy density

$$Q(T, \partial_t) = \tfrac{1}{4}Q(L + \underline{L}, aL + a^{-1}\underline{L})$$
$$= \tfrac{1}{4}[a(L\phi)^2 + a^{-1}(\underline{L}\phi)^2 + (a + a^{-1})|\nabla\!\!\!/\phi|^2].$$

Taking $X = e^a T$ as a multiplier with $a = a(u)$, we get from the weight additional interior terms

$$-a'(u)Q(L, T) = -\frac{a'(u)}{2}[(L\phi)^2 + |\nabla\!\!\!/ \phi|^2].$$

We choose $a'(s) = A\langle s\rangle^{-1-\epsilon}$ with A big enough, and finish the proof exactly as before (see [5] for details). ◇

Variations on this theme of improved standard inequality also appear in [30] and [41] (see chapter 9 for a statement).

4. Maxwell equations in the general case Let us work again with a null frame associated with one optical function, but consider now the Maxwell equations. Instead of defining the electric and magnetic fields as in the flat case (which means six functions altogether), let us define the six components of F in our null frame $(e_1, e_2, \underline{L}, L)$ as

$$\alpha_a = F(e_a, L),\ \underline{\alpha}_a = F(e_a, \underline{L}),\ \rho = \tfrac{1}{2}F(\underline{L}, L),\ \sigma = F(e_1, e_2).$$

Note that α and $\underline{\alpha}$ are 1-forms on the (nonstandard) 2-spheres, and that σ does not depend on the chosen orthonormal frames on these spheres, since for another frame

$$\tilde{e}_1 = (\cos\theta)e_1 - (\sin\theta)e_2,\ \tilde{e}_2 = (\sin\theta)e_1 + (\cos\theta)e_2,$$

we would find

$$\tilde{\sigma} = F(\tilde{e}_1, \tilde{e}_2) = -(\sin^2\theta)F(e_2, e_1) + (\cos^2\theta)F(e_1, e_2) = F(e_1, e_2) = \sigma.$$

First, we note that, computing the double trace $|F|^2$ in our null frame and taking into account the symmetries of F,

$$|F|^2 = 2F_{12}^2 - 2F_{13}F_{14} - 2F_{23}F_{24} - \tfrac{1}{2}F_{34}^2 = 2\sigma^2 - 2\rho^2 - 2\alpha\underline{\alpha}.$$

From the definition of the energy–momentum tensor in this case, we can write

$$Q(X, Y) = 2F(X, \partial_\gamma)F(Y, \partial^\gamma) - \tfrac{1}{2}\langle X, Y\rangle|F|^2,$$

and we observe that the first term is a trace, which can be computed in any basis with its dual basis. For instance,

$$\begin{aligned} F(X, \partial_\gamma)F(Y, \partial^\gamma) = {} & F(X, e_1)F(Y, e_1) + F(X, e_2)F(Y, e_2) \\ & - \tfrac{1}{2}F(X, \underline{L})F(Y, L) - \tfrac{1}{2}F(X, L)F(Y, \underline{L}). \end{aligned}$$

For the components of Q, we thus find

$$Q(L, L) = 2|\alpha|^2,\ Q(\underline{L}, \underline{L}) = 2|\underline{\alpha}|^2,\ Q(L, \underline{L}) = 2(\rho^2 + \sigma^2),$$
$$Q(L, e_1) = 2(-\sigma\alpha_2 + \rho\alpha_1),\ Q(\underline{L}, e_1) = 2(-\sigma\underline{\alpha}_2 - \rho\underline{\alpha}_1),$$
$$Q(e_1, e_1) = \sigma^2 + \rho^2 - \alpha_1\underline{\alpha}_1 + \alpha_2\underline{\alpha}_2,\ Q(e_1, e_2) = -(\alpha_1\underline{\alpha}_2 + \alpha_2\underline{\alpha}_1),$$

and similarly for the other components.

Finally, define the energy of F at time T to be

$$E(T) = \frac{1}{2}\int_{\Sigma_T} \{a|\alpha|^2 + a^{-1}|\underline{\alpha}|^2 + (a + a^{-1})(\rho^2 + \sigma^2)\}dv.$$

The following theorem is completely analogous to the corresponding theorem for $\Box$.

Theorem *Let $T = \frac{1}{2}(L + \underline{L})$ and $\pi = {}^{(T)}\pi$. Assume that, for some $\epsilon > 0$, the components of π satisfy the following estimates:*

(i) $\langle u\rangle^{1+\epsilon} a(\pi_{11}^2 + \pi_{12}^2 + \pi_{22}^2 + \pi_{1L}^2 + \pi_{2L}^2) \in L_t^1 L_x^\infty$,
(ii) $\langle u\rangle^{1+\epsilon}(|\pi_{11}| + |\pi_{22}| + |\pi_{1\underline{L}}| + |\pi_{2\underline{L}}| + |\underline{\omega}|) \in L_{x,t}^\infty$.

Then, for some constant $C = C_\epsilon$, for all solutions F of the Maxwell equations and all T,

$$E(T) + \int_{0\le t\le T} \langle u\rangle^{-1-\epsilon}(|\alpha|^2 + \rho^2 + \sigma^2)dtdv \le CE(0).$$

In particular, under these assumptions, the good components of F are α, ρ, σ.

The proof of this theorem is practically the same as for the case of $\Box$ in section 5.3. We use the multiplier $X = e^a T$ with $a = a(u)$. This gives additional terms in $Q\pi$:

$$a'(u)Q(L, L + \underline{L}) = 2a'(u)(|\alpha|^2 + \rho^2 + \sigma^2),$$

and identifies the good components. The corresponding energy density is

$$Q(X, \partial_t) = \frac{1}{2}e^a Q(T, aL + a^{-1}\underline{L}),$$

justifying the definition of E. The rest of the proof is the same as that in section 5.3 (see [5] for details). ◇

5.4 Conformal inequalities

1. In the flat case, this quite miraculous inequality (see [9], [24]) is obtained using the multiplier

$$K_0 = (r^2 + t^2)\partial_t + 2rt\partial_r.$$

Recall that K_0 is a timelike conformal Killing field. The question is now: for a general metric g (close to the flat case or more general), how should a substitute for K_0 be chosen? To motivate the answer, let us go back once more to the flat case, and set $u = t - r$, $\underline{u} = t + r$. We easily check the formula

$$S = t\partial_t + r\partial_r = \tfrac{1}{2}(\underline{u}L + u\underline{L}),\ K_0 = \tfrac{1}{2}(\underline{u}^2 L + u^2 \underline{L}).$$

Suppose now that we have a general split metric g and an optical function u, and that we work in the associated null frame as explained in section 2.3. Following the presentation of [28], by analogy with the flat case, define $\underline{u} = 2t - u$ (though terribly ugly, this formula will do), and set

$$K_0 = \tfrac{1}{2}(\underline{u}^2 L + u^2 \underline{L}).$$

2. For a certain function Ω, to be chosen later, let us set

$$\pi = {}^{(K_0)}\pi,\ \tilde{\pi} = \pi - \Omega g.$$

Similarly to what we did in chapter 4, we can modify the key formula for the energy inequality by writing

$$D^\alpha \tilde{P}_\alpha = \frac{1}{2} Q_{\alpha\beta}\tilde{\pi}^{\alpha\beta} - \frac{1}{4}\phi^2 \Box\Omega + \left(K_0\phi + \frac{\Omega}{2}\phi\right)(\Box\phi),$$

with the modified $\tilde{P}$

$$\tilde{P}_\alpha = Q_{\alpha\beta}K_0^\beta + \frac{\Omega}{2}\phi\partial_\alpha\phi - \frac{1}{4}\phi^2\partial_\alpha\Omega.$$

We choose $\Omega = 4t$, which is the value for the flat case. By integrating in a slab $\{0 \le t \le T\}$, we then obtain the **conformal energy** at time T

$$\begin{aligned}\tilde{E}(T) = \tfrac{1}{4}\int_{\Sigma_T} &\{a\underline{u}^2(L\phi)^2 + a^{-1}u^2(\underline{L}\phi)^2 \\ &+ (a^{-1}u^2 + a\underline{u}^2)|\nabla\!\!\!/\phi|^2 + 8t\phi\partial_t\phi - 4\phi^2\}dv.\end{aligned}$$

Recall here the notation

$$a = (\partial_t u)^{-1},\ N = -(\partial_t u)^{-1} g^{ij}\partial_i u \partial_j.$$

We prove now the positivity of the conformal energy.

Theorem *Assume* $|a-1| \leq 0.1$ *and* $(\text{div}\, N)(\underline{u}-u) = 4+\epsilon$ *with* $|\epsilon|$ *small enough. Then, for some constant* $C > 0$,

$$\tilde{E}(T) \geq C \int_{\Sigma_T} \{\underline{u}^2 (L\phi)^2 + (u^2+\underline{u}^2)|\nabla\!\!\!/ \phi|^2 + u^2(\underline{L}\phi)^2 + \phi^2\} dv.$$

This theorem identifies $L\phi$ and $\nabla\!\!\!/ \phi$ as the good derivatives of ϕ. The additional control of $u\underline{L}\phi$ is not useless (as we shall see in section 6.1). It is a weak form of a control of all $Z\phi$ (Z denoting a Lorentz field as usual), since

$$\underline{u}L + u\underline{L} = 2S,\ \underline{u}L - u\underline{L} = 2\sum \omega_i H_i.$$

Once again, note that our choice of the multiplier K_0 is coherent with the "key idea," since

$$\begin{aligned}\tilde{\pi}_{LL} = \pi_{LL} &= 2\langle D_L K_0, L\rangle = \langle D_L(\underline{u}^2 L + u^2 \underline{L}), L\rangle \\ &= L(\underline{u}^2)\langle L, L\rangle + \underline{u}^2 \langle D_L L, L\rangle + L(u^2)\langle \underline{L}, L\rangle - u^2\langle D_L L, \underline{L}\rangle = 0.\end{aligned}$$

Of course, the theorem provides only an analysis on the conformal energy; it should be complemented with a thorough analysis of the interior terms $\int_{\mathcal{D}} Q\pi\, dV$ in the spirit of section 4.5. We refer the reader to [28] or to [24] (in a nongeometric setting) for such an analysis.

Proof The idea of the proof is to transform the term $t\partial_t\phi$ of $\tilde{E}$, using two auxiliary fields S and $\underline{S}$, into a term involving a tangential derivative, for which we can perform an integration by parts on Σ_T.

(a) We set

$$S = \tfrac{1}{2}(a\underline{u}L + a^{-1}u\underline{L}),\ \underline{S} = \tfrac{1}{2}(a\underline{u}L - a^{-1}u\underline{L}).$$

Then

$$t\partial_t\phi = S\phi - \frac{1}{2}(\underline{u}-u)N\phi,\ t\partial_t\phi = \frac{2t}{\underline{u}-u}\underline{S}\phi - \frac{2t^2}{\underline{u}-u}N\phi.$$

For some number $0 < \lambda < 1$, to be determined, we split $t\partial_t\phi = \lambda t\partial_t\phi + (1-\lambda)t\partial_t\phi$ and use the expression with S for the first term, and the expression with $\underline{S}$ for the second.

(b) To integrate by parts in Σ_T, we need a formula. For any field X and any two functions f, h, we have

$$\text{div}\,(fgX) = fg\,\text{div}\, X + X(fg).$$

We will use this with $X = N$, noting that

$$\nu \equiv \operatorname{div} N = \sum \langle D_a N, e_a \rangle = \langle D_a(aL - T), e_a \rangle = a\operatorname{tr}\chi - k_a^a.$$

(c) Using the divergence formula of (b), we obtain

$$\begin{aligned}\int t\phi\partial_t\phi dv &= \int \phi S\phi dv - \tfrac{1}{4}\int (\underline{u} - u)N(\phi^2)dv \\ &= \int \phi S\phi dv + \tfrac{1}{4}\int [N(\underline{u} - u) + \nu(\underline{u} - u)]\phi^2 dv.\end{aligned}$$

Similarly, we get

$$\int t\phi\partial_t\phi dv = \int \frac{2t}{\underline{u} - u}\phi \underline{S}\phi dv + \int \left[N\left(\frac{t^2}{\underline{u} - u}\right) + \nu\frac{t^2}{\underline{u} - u}\right]\phi^2 dv.$$

Now, using the assumptions of the theorem, we find for the ϕ^2-coefficient in the first expression

$$\begin{aligned}N(\underline{u} - u) = 2N(t - u) = -2Nu = 2\langle N, L\rangle = 2,\ N(\underline{u} - u) + \nu(\underline{u} - u) \\ = 6 + \epsilon,\end{aligned}$$

and for the ϕ^2-coefficient in the second

$$N\left(\frac{t^2}{\underline{u} - u}\right) + \nu\frac{t^2}{\underline{u} - u} = \frac{t^2}{(\underline{u} - u)^2}[-2 + \nu(\underline{u} - u)] = (2 + \epsilon)\frac{t^2}{(\underline{u} - u)^2}.$$

(d) We now observe

$$a^2\underline{u}^2(L\phi)^2 + a^{-2}u^2(\underline{L}\phi)^2 = 2[(S\phi)^2 + (\underline{S}\phi)^2].$$

Since $|a - 1| \leq 0.1$, $d \leq a$, and $d \leq a^{-1}$ for $d = 0.9$, the integrand of $4\tilde{E}$ is bigger than $(a^{-1}u^2 + a\underline{u}^2)|\nabla\!\!\!/\phi|^2 + F$, with

$$F = 2d[(S\phi)^2 + (\underline{S}\phi)^2] + 8\lambda t\phi\partial_t\phi + 8(1 - \lambda)t\phi\partial_t\phi - 4\phi^2.$$

Integrating on Σ_T, we get

$$\begin{aligned}\int F dv = \int \Bigg\{ & 2d\left(S\phi + 2\lambda\frac{\phi}{d}\right)^2 \\ & + \phi^2\left[(12 + 2\epsilon)\lambda - 4 - 8\frac{\lambda^2}{d}\right] + 2d\left(\underline{S}\phi + 4t\frac{1 - \lambda}{d(\underline{u} - u)}\phi\right)^2 \\ & + \frac{t^2}{(\underline{u} - u)^2}\phi^2\left[8(1 - \lambda)(2 + \epsilon) - 32\frac{(1 - \lambda)^2}{d^2}\right]\Bigg\} dv.\end{aligned}$$

Taking $\epsilon = 0$, we need to impose the conditions

$$\lambda > 1 - \frac{d}{2}, \; 2\frac{\lambda^2}{d} - 3\lambda + 1 < 0.$$

Since $d = 0.9$, this reads

$$\lambda > 0.55, \; 0.6 < \lambda < 0.75,$$

which is certainly possible. Thus, choosing λ is also possible for $|\epsilon|$ small enough. $\diamond$

6

Pointwise estimates and commutations

In the previous chapters, we put emphasis on energy estimates, since nothing is possible without them: they provide the basic control of the solutions, and allow one to identify the "good components," as explained in chapter 5. However, for the sake of completeness and for applications to nonlinear equations, one generally needs more than (weighted) L^2 estimates: one also needs pointwise estimates, displaying the rate of decay and the qualitative behavior of the solutions.

The basic tool to use here is the **Klainerman inequality**

$$(1+|t|+r)^2(1+|r-t|)|v(x,t)|^2 \leq C \sum ||Z^k v(\cdot,t)||^2_{L^2_x},$$

where the fields Z are the standard Lorentz fields $Z = \partial_\alpha,\ S,\ R_i,\ H_i$ and the sum is extended to all products Z^k of k such fields, $k \leq 2$ (see, for instance, [24] for a proof). The "**Klainerman method**" is the fundamental strategy to obtain pointwise estimates. It consists of the following three steps:

Step 1 Prove an energy inequality for $\Box$.

Step 2 Commute products Z^k of Z fields with $\Box$ to obtain equations $\Box Z^k\phi = f_k$ $(k \leq N)$, and apply the energy inequality to these equations.

Step 3 Use the Klainerman inequality to obtain the pointwise qualitative behavior of $\nabla Z^l \phi$ $(l \leq N-2)$.

The only problem with this strategy is that, in general, there is no reason why the *standard* Lorentz fields Z should commute with $\Box = \Box_g$. Hence there are three possibilities:

(1) We use modified Z fields but do not commute them with $\Box$. This is made possible by using (generalized) conformal inequalities, which, as we saw in section 5.4, directly yield a bound for the L^2_x norms of

$$\underline{u}L\phi,\ u\underline{L}\phi,\ \underline{u}\,\nabla\!\!\!/\phi.$$

These fields can be considered good substitutes for the Z fields. In this case, however, we control only *one* Z field, and not *products* of such fields! We will see that this is enough to get some qualitative information, but this information is not as good as that from the Klainerman inequality.

(2) We change the *standard* Z into *modified* fields $\tilde{Z}$ which commute better with $\Box$. This is a rather difficult geometric construction, and we will see two aspects of it.

(3) We nevertheless use the standard Lorentz fields Z. Rather unexpectedly, this approach turns out to be efficient in many nonlinear *perturbation* problems. We postpone this discussion until the last chapter.

6.1 Pointwise decay and conformal inequalities

We saw in section 5.4 the expression for the modified energy $\tilde{E}$ which arises when establishing a conformal inequality. If we can bound $\tilde{E}$, we can bound in particular, the spatial L^2 norms of $r \not\nabla \phi$, $\underline{u} L\phi$, and $u\underline{L}\phi$.

The article [28] is written using a null frame associated with one optical function u, as explained in chapter 1. The strategy of [28] is the following:

- One does not try to commute the Z fields with $\Box$; one commutes only ∂_t, or more precisely, $T_0 = \frac{1}{2}(L + \underline{L})$. We have already seen in section 5.3 that ${}^{(T_0)}\pi_{LL} = 0$, which, as we shall see in the next section, is a condition ensuring cancellation of the bad terms in the commutator $[T_0, \Box]$. Using the equation $\Box u = f$, the control of the T_0 derivatives yields a control of all derivatives; for instance, in the flat case,

$$||\partial^2\phi||_{L^2} \leq ||\Delta\phi||_{L^2} \leq (||\partial_t^2 u||_{L^2} + ||\Box u||_{L^2}).$$

- One defines a higher order energy $\tilde{E}_{k+1}$ by the formula

$$\tilde{E}_{k+1} = \sum_{|\alpha|\leq k} \tilde{E}(\partial^\alpha \phi).$$

The following proposition gives the required pointwise bound.

Proposition *Let ϕ be smooth and sufficiently decaying as $|x| \to +\infty$. For $p > 2$,*

$$|\partial\phi(x,t)| \leq C_p(1+t)^{-\frac{2}{p}}\tilde{E}_3(t).$$

The drawback of this strategy is obviously that, even in the flat case, it can never give the good decay rate t^{-1}, since $p > 2$. However, it has the advantage of commuting only "ordinary" derivatives with $\Box$, instead of Z fields.

To see the full strength of the conformal energy, we outline briefly the proof of the proposition in the flat case. First, we admit the following functional analysis lemma.

Lemma *For any smooth function v on $\mathbf{R}^3_x$ sufficiently decaying at infinity, and $p > 2$, $s \geq 5/2 - 3/p$,*

$$|v(x)| \leq C|x|^{-\frac{2}{p}}(||r\nabla\!\!\!/ v||_{H^s} + ||v||_{H^s}).$$

If we consider a solution $\phi(x, t)$ of $\Box\phi = 0$, this lemma yields the inequality of the proposition only for $|x| \geq \frac{t}{2}$. For $|x| \leq \frac{t}{2}$, we just note that

$$\tilde{E} \geq Ct^2 \int_{|x|\leq\frac{t}{2}} |\partial\phi|^2 dx.$$

This is where the term $u\underline{L}\phi$ is used. The control of $\underline{u}L\phi$, $u\underline{L}\phi$, $r\nabla\!\!\!/\phi$ yields in fact the control of the hyperbolic rotations $H_i\phi$, since

$$H_0 \equiv \sum \omega_i H_i = t\partial_r + r\partial_t = \tfrac{1}{2}(\underline{u}L - u\underline{L}),$$

$$H_i = \omega_i H_0 + t(\partial_i - \omega_i\partial_r),\ \partial_i - \omega_i\partial_r = -\left(\omega \wedge \frac{R}{r}\right)_i .$$

6.2 Commuting fields in the scalar case

We establish a general commutation formula.

Theorem (Commutation formula) *For any field X with deformation tensor $\pi = {}^{(X)}\pi$,*

$$[\Box, X]\phi = \pi^{\alpha\beta}\nabla^2\phi_{\alpha\beta} + D_\alpha\pi^{\alpha\beta}\partial_\beta\phi - \tfrac{1}{2}\partial^\alpha(tr\,\pi)\partial_\alpha\phi.$$

In particular, X is a Killing field if and only if $[\Box, X] = 0$; in practice this is an easy way to identify a Killing field. For instance, if g is the Kerr metric, we see immediately that ∂_t and ∂_ϕ commute with $\Box$. If X is conformal Killing, that is, $\pi = \lambda g$, we have $D^\alpha\pi_{\alpha\beta} = \partial_\beta\lambda$, hence

$$[\Box, X]\phi = \lambda\Box\phi - \partial^\alpha\lambda\partial_\alpha\phi.$$

For the flat metric, for instance, we get $[\Box, S] = 2\Box$, but $[\Box, K_\mu] \neq -4x_\mu\Box$. We say that S commutes well with $\Box$, since if $\Box\phi = f$ is known, so is $\Box S\phi = Sf + 2f$; but this is not the case for K_μ. Note, however, the formula

$$\Box(K_0 + 2t)\phi = (K_0 + 6t)\Box\phi.$$

Proof We give here a (pedestrian) self-contained proof of the theorem, using the formula for $\pi^{\alpha\beta}$ given in chapter 4.

(a) We write $\Box\phi = \partial^\alpha\partial_\alpha\phi - (D^\alpha\partial_\alpha)\phi$, hence

$$X\Box\phi = [X, \partial^\alpha]\partial_\alpha\phi + \partial^\alpha[X, \partial_\alpha]\phi - [X, D^\alpha\partial_\alpha]\phi + \Box X\phi.$$

Now

$$[X, \partial_\alpha] = -(\partial_\alpha X^\mu)\partial_\mu, \ [X, \partial^\alpha] = X(g^{\alpha\beta})\partial_\beta - (\partial^\alpha X^\mu)\partial_\mu.$$

Gathering the terms, we get, with a first order differential term E,

$$[\Box, X]\phi = \pi^{\alpha\beta}\nabla^2\phi_{\alpha\beta} + E,$$
$$E = \pi^{\alpha\beta}(D_\alpha\partial_\beta)\phi + (\partial^\alpha g_{\alpha\gamma})(\partial^\gamma X^\beta)\partial_\beta\phi + \partial_\alpha(\partial^\alpha X^\beta)\partial_\beta\phi + [X, D^\alpha\partial_\alpha]\phi.$$

(b) From the definition of $D\pi$, we get

$$\begin{aligned} D^\alpha\pi_{\alpha\beta} &= \langle D^\alpha D_\alpha X, \partial_\beta\rangle + \langle D^\alpha D_\beta X, \partial_\alpha\rangle \\ &\quad - \langle D_{D^\alpha\partial_\alpha} X, \partial_\beta\rangle - \langle D_{D^\alpha\partial_\beta} X, \partial_\alpha\rangle. \end{aligned}$$

Also,

$$\tfrac{1}{2}\partial_\beta tr\,\pi = \langle D_\beta D_\alpha X, \partial^\alpha\rangle + \langle D_\alpha X, D_\beta\partial^\alpha\rangle.$$

Hence

$$D^\alpha\pi_{\alpha\beta} - \tfrac{1}{2}\partial_\beta tr\,\pi = I + II + III,$$

where

$$\begin{aligned} III &= \langle D_{[\partial^\alpha, X]}\partial_\alpha, \partial_\beta\rangle + \langle D^\alpha[\partial_\alpha, X], \partial_\beta\rangle \\ &\quad + \langle [X, D^\alpha\partial_\alpha], \partial_\beta\rangle - \pi(D^\alpha\partial_\beta, \partial_\alpha) - (\partial_\beta g^{\alpha\gamma})\langle D_\alpha X, \partial_\gamma\rangle. \end{aligned}$$

Here, we have introduced on purpose the terms

$$I = \langle D_\alpha D_\beta X - D_\beta D_\alpha X, \partial^\alpha\rangle,$$
$$II = \langle D^\alpha D_X\partial_\alpha - D_X D^\alpha\partial_\alpha - D_{[\partial^\alpha, X]}\partial_\alpha, \partial_\beta\rangle.$$

These terms are "curvature terms," denoted

$$I = R^\alpha_{X\alpha\beta}, \ II = R^\alpha_{\beta\alpha X}.$$

For simplicity, we chose to introduce the curvature tensor R only in the next chapter, so we have to admit at this point the symmetries of this tensor:

$$R_{\alpha\beta\gamma\delta} = -R_{\beta\alpha\gamma\delta} = R_{\gamma\delta\alpha\beta}.$$

This being admitted, we obtain $I + II = 0$, and

$$III = \pi^{\alpha\gamma}\langle D_\gamma \partial_\alpha, \partial_\beta\rangle - \pi^{\alpha\gamma}\langle D_\alpha \partial_\beta, \partial_\gamma\rangle$$
$$+ \partial^\alpha(\partial_\alpha X^\gamma) g_{\gamma\beta} + \langle [X, D^\alpha \partial_\alpha], \partial_\beta\rangle - (\partial_\beta g^{\alpha\gamma})\langle D_\alpha X, \partial_\gamma\rangle.$$

(c) Finally,

$$(D^\alpha \pi_{\alpha\beta} - \tfrac{1}{2}\partial_\beta \mathrm{tr}\,\pi)\partial^\beta\phi - E = \tfrac{1}{2}(\nabla\phi)(g_{\alpha\beta})\pi^{\alpha\beta} - \tfrac{1}{2}(\nabla\phi)(g_{\alpha\gamma}\pi^{\alpha\gamma}) = 0,$$

since

$$\pi^{\alpha\gamma}\langle D_\alpha \partial_\beta, \partial_\gamma\rangle = \pi^{\alpha\gamma}\langle D_\beta \partial_\alpha, \partial_\gamma\rangle = \tfrac{1}{2}\pi^{\alpha\gamma}\partial_\beta g_{\alpha\gamma}. \qquad \diamond$$

Like all such formulas, this commutation formula has the advantage of being written as multiple traces: the term $\pi^{\alpha\beta}\nabla^2\phi_{\alpha\beta}$ is a double trace, analogous to the term $\pi^{\alpha\beta}Q_{\alpha\beta}$ of the energy inequalities; the other terms are

$$D_\alpha \pi^{\alpha\beta}\partial_\beta\phi - \tfrac{1}{2}\partial^\beta(\mathrm{tr}\,\pi)\partial_\beta\phi = D_\alpha \pi^\alpha_{\nabla\phi} - \tfrac{1}{2}(\nabla\phi)(\mathrm{tr}\,\pi).$$

6.3 Modified Lorentz fields

In practice, we cannot expect to find X commuting exactly with $\Box$, so we choose modified Lorentz fields X which look close to the standard Lorentz fields *and* can be expressed in the null frame with which we are working.

1. Good commutation condition If one is mainly concerned with decay at infinity, one can use the concept of the "good derivative" to construct modified fields. In chapter 4, when looking for a good multiplier X to obtain an energy inequality, we found the condition ${}^{(X)}\pi_{LL} = 0$; this condition ensured that all of the interior terms $\pi^{\alpha\beta}Q_{\alpha\beta}$ would contain at least one "good derivative" e_1, e_2, or L. Similarly, we can sketch what could be a "good commuting field" X for $\Box$. This is a field X such that the higher order terms $\pi^{\alpha\beta}\nabla^2\phi_{\alpha\beta}$ (given by the theorem in section 6.2) involve only good derivatives of ϕ, that is, second order derivatives containing at least one good derivative. Since, with this definition, the only bad second order derivative is $\underline{L}^2\phi$, the required condition is again ${}^{(X)}\pi_{LL} = 0$.

Good commutation condition ${}^{(X)}\pi_{LL} = 0$.

2. The difficulty with the hyperbolic rotations Consider the equation $-\partial_t^2 + c^2\Delta_x$ instead of the standard $\Box = -\partial_t^2 + \Delta_x$. For the hyperbolic rotations, we now have to take $H_i = \frac{x^i}{c}\partial_t + ct\partial_i$. In other words, H_i depends on the speed

c, while S and R_i do not. It turns out that, in more general situations, good substitutes for H_i are not known.

This being admitted, one has to explain how to replace the Klainerman inequality when the fields H_i are missing. The idea is to use only the fields $\Gamma = \partial_\alpha,\ S,\ R$, and the operator $\Box$ itself. For the flat case, one can prove the following inequalities (see [38]), which can be viewed as a substitute for the Klainerman inequality.

Proposition *Let $E(\phi)$ be the standard energy, and define a higher order energy E_{k+1} by*

$$E_{k+1}(\phi) = \sum_{|\alpha|\leq k} E(\Gamma^\alpha \phi),\ \Gamma = \{\partial_\alpha,\ S,\ R\}.$$

Then, ∂_i $(i = 1, 2, 3)$ being the spatial derivatives,

$$(1+r)\langle t-r\rangle|\partial_i \nabla\phi| \leq C(E_4^{\frac{1}{2}} + t||\Box\phi||_{L^2}).$$

It turns out that this type of inequality is also available for more general situations than the flat case (see, for instance, [6]). Thus the Klainerman method can be extended to nonflat geometric situations, dropping the H_i and using appropriately defined S and $\tilde{R}_i$.

3. The fully geometric approach This is, for instance, the approach of [29], where the null frame is associated with two optical functions u and $\underline{u}$ as explained in chapter 2. In such a geometric framework, one is looking for substitutes for the standard Lorentz fields $S,\ R_i,\ H_i$. We have already mentioned that it is necessary and possible to forget about the hyperbolic rotations H_i. The generalization for S follows easily from the formula $S = \frac{1}{2}(u\underline{L} + \underline{u}L)$ in chapter 5. For R_i, we take fields tangent to the (nonstandard) 2-spheres of the foliation. The actual construction in [29] is rather delicate, and we only sketch it here. First, we consider in $\Sigma_0 = \{t = 0\}$ a specific sphere foliation, with unitary normal field N. The flow of N and the asymptotic properties of the metric at infinity allow one to pull back to a given sphere the standard rotations R_i at infinity (which are homogeneous of degree 0). Once this is done, we push forward these rotation fields by the flow of L along an outgoing cone. In this way, we obtain rotations ${}^iO$ satisfying good commutation relations

$$[{}^iO, {}^jO] = \epsilon_{ijk}{}^kO,\ [L, {}^iO] = 0.$$

The drawback of this definition is, of course, its global and nonexplicit character.

4. A simplified approach Suppose for instance that we choose a quasiradial frame and that our assumptions on g allow us to identify e_1, e_2, and L as the good derivatives (see chapter 4). The idea for constructing a good commuting modified field $\tilde{Z}$ is to try $\tilde{Z} = Z + aT$, since the other terms in the perturbation of Z would involve only good derivatives and would probably play a negligible role (note that here we take T rather than $\underline{L}$ since T has smooth coefficients everywhere). Then

$$^{(\tilde{Z})}\pi_{LL} = {}^{(Z)}\pi_{LL} + a^{(T)}\pi_{LL} - 2La = 2\langle [L, Z], L\rangle - 2a\frac{Tc}{c} - 2La.$$

For the reasons explained above, we forget about H_i and will use and modify only $Z = R_i$ and $Z = S$. Since $[R_i, \partial_r] = 0$, $[S, R_i] = 0$,

$$[R, T] = [R, \partial_t] + \cdots R,\ [R, N] = \frac{Rc}{c}N + \cdots R,\ [S, T] = [S, \partial_t] + \cdots R,$$
$$[S, N] = \frac{Sc}{c}N + c[S, \partial_r] + \cdots R,\ [S, L] = -L + \frac{Sc}{c}N + \cdots R,$$

we get

$$\langle [L, R_i], L\rangle = -\frac{Rc}{c},\ \langle [L, S], L\rangle = -\frac{Sc}{c}.$$

Finally, we define $a = {}^Z a$ by

$$La + a\frac{Tc}{c} + \frac{Zc}{c} = 0,$$

and use the fields

$$\tilde{Z} = Z + {}^Z aT,$$

which satisfy the "good commutation" relation. Of course, this causes a certain number of technical difficulties: we need to control a simultaneously with the solution ϕ to obtain a pointwise estimate from a bound of $\nabla \tilde{Z}^k \phi$ in L^2 norm, etc. The advantage by comparison with the fully geometric approach is more simplicity in the computations. We refer the reader to [6] for details.

6.4 Commuting fields for Maxwell equations

In the scalar case of the previous section, our aim was to control $X\phi$, or more generally $X^k\phi$ for some collection of fields X, ϕ being a solution of $\Box\phi = 0$. For Maxwell equations, the unknown is a 2-form F, and XF does not make sense; it has to be replaced by $\mathcal{L}_X F$, the Lie derivative that we briefly discussed in chapter 4. Since the exterior differential operator d commutes with mappings, we have $[\mathcal{L}_X, d] = 0$. Thus, for a solution of Maxwell equations,

$$\mathcal{L}_X(dF) = 0 = d(\mathcal{L}_X F),$$
$$\mathcal{L}_X(d * F) = 0 = d(\mathcal{L}_X * F).$$

To compute the commutation defect coming from the second equation, one can use the following formula, where π stands for the deformation tensor of X:

$$\mathcal{L}_X * F_{\mu\nu} = *\mathcal{L}_X F_{\mu\nu} + *F_\mu^{\ \rho}\pi_{\rho\nu} + *F_\nu^{\ \rho}\pi_{\mu\rho} - \tfrac{1}{2}\mathrm{tr}\,\pi * F_{\mu\nu}.$$

Using this formula, one again obtains the set of Maxwell equations for $\mathcal{L}_X F$, perturbed by first order derivatives of F. Let us point out that there is an analogous formalism for the Bianchi identities that we will very briefly discuss in the last chapter. We refer the reader to [29] for more details about commutators in these cases.

7
Frames and curvature

In this chapter, we always assume for simplicity that the metric g is split,

$$g_{\alpha\beta}dx^\alpha dx^\beta = -dt^2 + g_{ij}dx^i dx^j,$$

that the coefficients g_{ij} and g^{ij} are bounded, and that, for some $C > 0$ and all points,

$$g_{ij}\xi^i\xi^j \geq C|\xi|^2.$$

We wish to examine in more detail how one can control the frame coefficients of a given null frame. At this point, we need to introduce the curvature tensor R.

7.1 The curvature tensor

1. The three fields X, Y, Z being given, the field $R(X, Y)Z$ is defined by the formula

$$R(X, Y)Z = D_X D_Y Z - D_Y D_X Z - D_{[X,Y]}Z.$$

Hence $R(X, Y)Z$ measures the commutation defect of D_X and D_Y, when applied to Z. The remarkable feature here is that this expression is *linear* in all three fields X, Y, Z. For instance,

$$R(fX, Y)Z = fD_X D_Y Z - D_Y(fD_X Z) - D_{f[X,Y]-(Yf)X}Z = fR(X, Y)Z,$$

$$\begin{aligned} R(X, Y)(fZ) &= D_X((Yf)Z + fD_Y Z) - D_Y((Xf)Z + fD_X Z) - D_{[X,Y]}(fZ) \\ &= (XYf)Z + (Yf)D_X Z + (Xf)D_Y Z - (YXf)Z \\ &\quad -(Xf)D_Y Z - (Yf)D_X Z - ([X, Y]f)Z + fR(X, Y)Z = fR(X, Y)Z. \end{aligned}$$

Note also that, by construction,

$$R(Y, X)Z = -R(X, Y)Z.$$

2. The **curvature tensor** is just

$$R(W, Z, X, Y) = \langle R(X, Y)Z, W\rangle.$$

The point of introducing all four arguments X, Y, Z, W lies in the remarkable symmetries of R:

$$R_{\alpha\beta\gamma\delta} = -R_{\beta\alpha\gamma\delta} = -R_{\alpha\beta\delta\gamma},$$
$$R_{\alpha\beta\gamma\delta} = R_{\gamma\delta\alpha\beta}.$$

The symmetry in $\gamma\delta$ has already been shown; the symmetry in $\alpha\beta$ is some "integration by parts" formula, since the derivatives acting on $Z = \partial_\beta$ are transferred to act on $W = \partial_\alpha$. The symmetry of the second line is more mysterious, and we refer the reader to standard textbooks for its proof (see, for instance, [22]). Finally, we also have the "circular permutation" formula

$$R_{\alpha\beta\gamma\delta} + R_{\alpha\delta\beta\gamma} + R_{\alpha\gamma\delta\beta} = 0.$$

Using the definition and the Christoffel symbols, one can write down an explicit formula in local coordinates for the components of R:

$$R_{\delta\alpha\beta\gamma} = \partial_\beta\Gamma_{\delta\gamma\alpha} - \partial_\gamma\Gamma_{\delta\alpha\beta} - g_{\mu\nu}\Gamma^\mu_{\alpha\gamma}\Gamma^\nu_{\beta\delta} + g_{\mu\nu}\Gamma^\mu_{\alpha\beta}\Gamma^\gamma_{\gamma\delta}.$$

To prove this, we just write

$$\begin{aligned} R_{\delta\alpha\beta\gamma} &= \langle D_\beta D_\gamma\partial_\alpha, \partial_\delta\rangle - \langle D_\gamma D_\beta\partial_\alpha, \partial_\delta\rangle \\ &= \partial_\beta\langle D_\gamma\partial_\alpha, \partial_\delta\rangle - \langle D_\gamma\partial_\alpha, D_\beta\partial_\delta\rangle - \partial_\gamma\langle D_\beta\partial_\alpha, \partial_\delta\rangle + \langle D_\beta\partial_\alpha, D_\gamma\partial_\delta\rangle. \end{aligned}$$

Thus R is expressed using second order derivatives of the metric g. Generally speaking, whenever a computation involves second order derivatives of g, one can expect R to appear; we will follow this path in many proofs in this chapter.

3. The **Ricci tensor** is a trace taken on R (because of the symmetries of R, there are not many traces to be taken)

$$Ric_{\mu\nu} = g^{\alpha\beta}R_{\mu\alpha\nu\beta} = R_{\mu\alpha\nu}^{\ \ \ \ \alpha},$$

and the **scalar curvature** is the trace of the Ricci tensor $R = Ric^\alpha_\alpha$. Note that, due to the symmetries of R, the Ricci tensor is symmetric

$$Ric_{\mu\nu} = g^{\alpha\beta}R_{\mu\alpha\nu\beta} = g^{\alpha\beta}R_{\nu\beta\mu\alpha} = g^{\beta\alpha}R_{\nu\beta\mu\alpha} = Ric_{\nu\mu}.$$

From the explicit formula for R we easily get in local coordinates

$$\begin{aligned} Ric_{\mu\nu} &= \tfrac{1}{2}\partial^\alpha \partial_\mu g_{\alpha\nu} + \tfrac{1}{2}\partial^\alpha \partial_\nu g_{\alpha\mu} - \tfrac{1}{2} g^{\alpha\beta}\partial^2_{\mu\nu} g_{\alpha\beta} \\ &\quad - \tfrac{1}{2} g^{\alpha\beta}\partial^2_{\alpha\beta} g_{\mu\nu} + g^{\alpha\beta} g_{\gamma\delta}(\Gamma^\gamma_{\mu\beta}\Gamma^\delta_{\alpha\nu} - \Gamma^\gamma_{\mu\nu}\Gamma^\delta_{\alpha\beta}). \end{aligned}$$

From now on, we just write $R_{\alpha\beta}$ for $Ric_{\alpha\beta}$.

7.2 Optical functions and curvature

1. New normalization of the null frame associated with an optical function From now on, to fit with [31] better, we are going to change slightly the choice of the null frame associated with the optical function u (see section 2.3). In this new normalization, we set $L' = -\nabla u$, and take, still noting $a = (\partial_t u)^{-1}$,

$$L = aL' = \partial_t + N, \ \underline{L} = \partial_t - N.$$

Thus, as usual,

$$\langle L, L\rangle = 0, \langle \underline{L}, \underline{L}\rangle = 0, \langle L, \underline{L}\rangle = -2.$$

This normalization is in some sense more natural than the previous one; its drawback is that L is no longer the gradient of some quantity. With these new definitions, the formulas for the frame coefficients have to be slightly modified. We give only the results, the proofs being analogous to that of section 3.4.

Theorem (Frame coefficients) *The frame coefficients are given by the following formulas:*

$$\begin{aligned} D_a L &= \chi_{ab} e_b - k_{aN} L, \ D_a \underline{L} = \underline{\chi}_{ab} e_b + k_{aN}\underline{L}, \\ D_L L &= -k_{NN} L, \ D_L \underline{L} = 2\underline{\eta}_a e_a + k_{NN}\underline{L}, \\ D_{\underline{L}} L &= 2\eta_a e_a + k_{NN} L, \ D_{\underline{L}}\underline{L} = 2\underline{\xi}_a e_a - k_{NN}\underline{L}, \\ D_L e_a &= \not{D}_L e_a + \underline{\eta}_a L, \ D_{\underline{L}} e_a = \not{D}_{\underline{L}} e_a + \eta_a \underline{L} + \underline{\xi}_a L, \\ D_b e_a &= \not{D}_b e_a + \tfrac{1}{2}\chi_{ab}\underline{L} + \tfrac{1}{2}\underline{\chi}_{ab} L. \end{aligned}$$

The formulas expressing $\underline{\chi}$, $\underline{\xi}$, $\underline{\eta}$ in terms of χ, ξ, η, given in section 3.4 also have to be slightly modified and now read

$$\underline{\chi}_{ab} = -\chi_{ab} - 2k_{ab},$$

$$\underline{\xi}_a = -\eta_a + k_{aN}, \ \underline{\eta}_a = -k_{aN}, \eta_a = \frac{e_a(a)}{a} + k_{aN}.$$

An important fact which will be useful later is the following.

Lemma *With the new normalization, the coordinates of L are bounded.*

Proof This is clear for the t-coordinate. Since $N = -a\partial^i u \partial_i$, the i-coordinate of L is $-g^{ij}\partial_j u/\partial_t u$. Taking into account the eikonal equation $(\partial_t u)^2 = g^{ij}\partial_i u \partial_j u$ and the assumptions on the metric, the claim is proved. ◇

2. Transport equation for u Let u be an optical function, that is, $\langle \nabla u, \nabla u \rangle = 0$. The transport equation satisfied by the derivatives of u is given in the following theorem.

Theorem *The second order derivatives of u are related to the curvature tensor through the transport equation*

$$D_L \nabla^2 u_{\alpha\beta} - a\nabla^2 u_{\alpha\gamma} \nabla^2 u^\gamma_\beta = a^{-1} R_{\beta L \alpha L}.$$

Proof Using the definitions, we get

$$D_L \nabla^2 u_{\alpha\beta} = L(\nabla^2 u_{\alpha\beta}) - \nabla^2 u(D_L \partial_\alpha, \partial_\beta) - \nabla^2 u(\partial_\alpha, D_L \partial_\beta),$$
$$L(\nabla^2 u_{\alpha\beta}) = L\langle D_\alpha \nabla u, \partial_\beta \rangle = \langle D_L D_\alpha \nabla u, \partial_\beta \rangle + \langle D_\alpha \nabla u, D_L \partial_\beta \rangle.$$

Since $D_L \nabla u = -a D_{\nabla u} \nabla u = 0$, using the symmetry of the Hessian, we obtain

$$\begin{aligned} D_L \nabla^2 u_{\alpha\beta} = & \langle D_L D_\alpha \nabla u, \partial_\beta \rangle - \langle D_\alpha D_L \nabla u, \partial_\beta \rangle \\ & - \langle D_{[L, \partial_\alpha]} \nabla u, \partial_\beta \rangle + \langle D_\beta \nabla u, [L, \partial_\alpha] \rangle - \langle D_\beta \nabla u, D_L \partial_\alpha \rangle. \end{aligned}$$

Taking into account $[L, \partial_\alpha] = D_L \partial_\alpha + D_\alpha(a\nabla u)$, the formula is proved. ◇

Recall from section 3.4 that the frame coefficients are just components of $\nabla^2 u$. It seems that this formula shows the necessity, at this stage, to introduce the curvature tensor in order to be able to control the frame coefficients.

The above formulas can be viewed as a system of differential equations in the unknowns $\nabla^2 u_{\alpha\beta}$, along the integral curves of L. Assuming that the components $R_{\alpha L \beta L}$ of the curvature tensor are given, we can use the above proposition to compute the components of $\nabla^2 u$, by solving the differential equations along L, with initial data on $\{t = 0\}$, on the t-axis, or elsewhere. To understand why this is not the best strategy, one has to imagine that the metric g is not smooth, and that we wish to pay the greatest attention to the regularity of the various objects at hand, counting derivatives, etc. In this context (which will be discussed briefly in the last chapter), the components of R have two derivatives fewer than g, and integrating along L does not gain anything.

In particular, consider the intrinsic components (see section 3.4)

$$\chi_{ab} = \langle D_a L, e_b \rangle = -a\langle D_a \nabla u, e_b \rangle = -a\nabla^2 u_{ab}.$$

It turns out that we have to split χ and $\underline{\chi}$ into their traces and their traceless parts

$$\operatorname{tr}\chi = \chi_a^a, \ \operatorname{tr}\underline{\chi} = \underline{\chi}_a^a, \ \chi = \hat{\chi} + \tfrac{1}{2}(\operatorname{tr}\chi)g, \ \underline{\chi} = \hat{\underline{\chi}} + \tfrac{1}{2}(\operatorname{tr}\underline{\chi})g.$$

The traces will be controlled by integration along L, while the traceless parts will be controlled through an elliptic system on the spheres of the foliation.

7.3 Transport equations

Theorem (Transport equations) *The quantities a and $\operatorname{tr}\chi$ satisfy*

$$La = -ak_{NN},$$
$$L(\operatorname{tr}\chi) + \tfrac{1}{2}(\operatorname{tr}\chi)^2 = -|\hat{\chi}|^2 - k_{NN}\operatorname{tr}\chi - R_{LL}.$$

Proof (a) We first prove the second formula. Since $\chi_{ab} = -a\nabla^2 u_{ab}$, $\operatorname{tr}\chi = -a\nabla^2 u_a^a$. In order to obtain $L(\operatorname{tr}\chi)$ from the transport equation on u, we observe that L commutes with the partial trace

$$L(\nabla^2 u_a^a) = D_L \nabla^2 u_a^a.$$

In fact,

$$L(\nabla^2 u_a^a) = D_L \nabla^2 u_a^a + 2\nabla^2 u(D_L e_a, e_a).$$

Now, since $\langle D_L e_a, e_a\rangle = 0$, $\langle D_L e_a, L\rangle = 0$ and $\langle D_L e_1, e_2\rangle = -\langle D_L e_2, e_1\rangle$, we have for some coefficients α, β_a,

$$D_L e_1 = \alpha e_2 + \beta_1 L, \ D_L e_2 = -\alpha e_1 + \beta_2 L.$$

Hence

$$\nabla^2 u(D_L e_1, e_1) + \nabla^2 u(D_L e_2, e_2) = \nabla^2 u(e_1, e_2)(\alpha - \alpha) + \beta_a \nabla^2 u(L, e_a).$$

Since $D_L \nabla u = 0$, $\nabla^2 u(L, e_a) = 0$ and the claim is proved. We thus obtain

$$L(\operatorname{tr}\chi) = \frac{La}{a}\operatorname{tr}\chi - aD_L\nabla^2 u_a^a.$$

From the transport equations on u, we immediately get

$$D_L \nabla^2 u_a^a = a\nabla^2 u_{a\gamma}\nabla^2 u_a^\gamma + a^{-1}\sum R_{aLaL}.$$

From the definition of the Ricci tensor R, using the symmetries of R,

$$R_{LL} = \sum R_{aLaL} - \tfrac{1}{2}R_{\underline{L}LLL} - \tfrac{1}{2}R_{LL\underline{L}L} = \sum R_{aLaL}.$$

As we have already observed that $\nabla^2 u_{La} = 0$, the trace in the above formula is just

$$\nabla^2 u_{a\gamma} \nabla^2 u_a^\gamma = \sum \nabla^2 u_{ab} \nabla^2 u_{ab} = a^{-2} |\chi|^2.$$

Summarizing, the formula is proved, since $|\chi|^2 = |\hat{\chi}|^2 + \frac{1}{2}(\mathrm{tr}\,\chi)^2$.

(b) To prove the first formula, we observe that

$$\langle D_L L, \underline{L} \rangle = -2\frac{La}{a},$$

since $D_L L = D_L(-a\nabla u) = (La/a)L$. On the other hand, since T and N are orthogonal unit vectors with $D_T T = 0$,

$$\langle D_L L, \underline{L} \rangle = \langle D_{T+N}(T+N), T-N \rangle = \langle D_T T, T-N \rangle \\ + \langle D_T N, T-N \rangle + \langle D_N T, T-N \rangle + \langle D_N N, T-N \rangle = 2k_{NN}.$$

◇

The point of these formulas is that they involve only k (first order derivatives of g), $\hat{\chi}$ (which will be separately controlled later by an elliptic system), and R_{LL}. We have to show now what is so special about R_{LL}.

Theorem (Special structure of R_{LL}) *Let $z = L^\nu g^{\alpha\beta} \partial_\beta g_{\alpha\nu} - \frac{1}{2} g^{\alpha\beta} L(g_{\alpha\beta})$. Then*

$$R_{LL} = Lz - \tfrac{1}{2} L^\mu L^\nu \Box g_{\mu\nu} + E,$$

where, for some constant C, $|E| \leq C|\partial g|^2$.

Proof The proof is by brute force, using the explicit formula given for $R_{\mu\nu}$. Observe first that the quadratic terms in Γ in the formula for $R_{\mu\nu}$ can be put into E. Next,

$$g^{\alpha\beta} \partial^2_{\alpha\beta} g_{\mu\nu} = \Box g_{\mu\nu} + E',$$

where E' can be put into E. We are left with the first three terms of the formula, which can be handled similarly, so we discuss only the first one, $g^{\alpha\beta} L^\mu L^\nu \partial^2_{\mu\beta} g_{\alpha\nu}$. We just write it

$$L(L^\nu g^{\alpha\beta} \partial_\beta g_{\alpha\nu}) - L(L^\nu) \partial^\alpha g_{\alpha\nu} - L^\nu L(g^{\alpha\beta}) \partial_\beta g_{\alpha\nu},$$

and observe that the last term can be put into E, while the first one enters into z. It remains to examine $L(L^\nu)$. Since $L = -a\nabla u$, $L^\nu = -a g^{\nu\mu} \partial_\mu u$. Hence

$$L(L^\nu) = \frac{La}{a} L^\nu + L(g^{\nu\mu}) \langle L, \partial_\mu \rangle + g^{\nu\mu} \langle L, D_L \partial_\mu \rangle.$$

Taking into account that $La/a = -k_{NN}$, all three terms are linear combinations, with bounded coefficients, of first order derivatives of g. Thus the term $L(L^\nu)\partial^\alpha g_{\alpha\nu}$ can be put into E, and this finishes the proof. ◇

7.4 Elliptic systems

These systems will control the traceless part of χ on one hand, and η on the other hand. Recall that χ and η are tensors on the spheres, so we will consider the induced connexion $\not{D}$ on the spheres to take their derivatives.

Recall that, if X or Y is not tangent to the spheres, we have defined $\not{D}_X Y$ as the orthogonal projection of $D_X Y$ on the spheres. This definition makes sense, since we saw in section 3.2 that if X and Y are tangent to the spheres, this projection is just the result of the induced connexion. We can extend this definition to tensors by the usual derivation formula

$$XT(Y, Z) = (\not{D}_X T)(Y, Z) + T(\not{D}_X Y, Z) + T(Y, \not{D}_X Z).$$

This extension will be used in the proofs.

1. Codazzi equation The elliptic system satisfied by $\hat{\chi}$ is given in the following theorem.

Theorem (Codazzi equation) *The tensor $\hat{\chi}$ satifies*

$$\mathrm{div}\, \hat{\chi}_a + \hat{\chi}_{ab} k_{bN} = \tfrac{1}{2}(e_a(\mathrm{tr}\, \chi) + k_{aN} \mathrm{tr}\, \chi) + R_{bLba}.$$

Recall that $\hat{\chi}$ is a symmetric 2-tensor on the sphere foliation, and div means the trace with respect to one argument:

$$\mathrm{div}\, \hat{\chi}_a = \not{D}_b \hat{\chi}_a^b.$$

As usual, one should be careful that we consider first $\not{D}_b \hat{\chi}$, and then take the ab component and sum. This is different from taking the divergence of the 1-form $\hat{\chi}(e_a, .)$.

Proof (a) To prove the Codazzi equation, we first prove

$$\not{D}_c \chi_{ab} - \not{D}_a \chi_{bc} = R_{bLca} - k_{cN} \chi_{ab} + k_{aN} \chi_{cb}.$$

This follows smoothly from the definitions; in fact,

$$\begin{aligned} \not{D}_c \chi_{ab} &= e_c(\chi_{ab}) - \chi(\not{D}_c e_a, e_b) - \chi(e_a, \not{D}_c e_b) \\ &= \langle D_c D_a L, e_b \rangle + \langle D_a L, D_c e_b \rangle - \langle D_b L, \not{D}_c e_a \rangle - \langle D_a L, \not{D}_c e_b \rangle. \end{aligned}$$

Since a second order derivative of L appears, we wish to introduce the missing terms to see a curvature term. Thus we write

$$\begin{aligned} \not{D}_c \chi_{ab} = R_{bLca} + \langle D_a D_c L, e_b \rangle + \langle D_{[e_c, e_a]} L, e_b \rangle \\ + \langle D_a L, D_c e_b - \not{D}_c e_b \rangle - \langle D_b L, \not{D}_c e_a \rangle. \end{aligned}$$

We transform terms according to the formulas

$$\begin{aligned} \langle D_a D_c L, e_b \rangle &= e_a(\chi_{bc}) - \langle D_c L, D_a e_b \rangle, \\ \langle D_{[e_c, e_a]} L, e_b \rangle &= \langle D_b L, [e_c, e_a] \rangle = \langle D_b L, \not{D}_c e_a - \not{D}_a e_c \rangle, \end{aligned}$$

and we get

$$\begin{aligned} \not{D}_c \chi_{ab} &= R_{bLca} + I + II, \\ I &= e_a(\chi_{bc}) - \langle D_b L, \not{D}_a e_c \rangle - \langle D_c L, \not{D}_a e_b \rangle, \\ II &= \langle D_a L, D_c e_b - \not{D}_c e_b \rangle - \langle D_c L, D_a e_b - \not{D}_a e_b \rangle. \end{aligned}$$

Since

$$\not{D}_a \chi_{bc} = e_a(\chi_{bc}) - \langle D_c L, \not{D}_a e_b \rangle - \langle D_b L, \not{D}_a e_c \rangle = I,$$

we are left to deal with the term II of the second line.

In section 3.2, in the case of a submanifold of codimension 1, we introduced the second form k, and proved the formula $D_X Y = \not{D}_X Y + k(X, Y)N$. Here, we are dealing with the submanifold $S_{t,u}$ of codimension 2, for which χ and $\underline{\chi}$ act as a pair of second forms. Just as in section 3.2, we can prove the formula

$$D_a e_b - \not{D}_a e_b = \tfrac{1}{2} \chi_{ab} \underline{L} + \tfrac{1}{2} \underline{\chi}_{ab} L.$$

In fact,

$$\langle D_a e_b, L \rangle = e_a \langle e_b, L \rangle - \langle e_b, D_a L \rangle = -\chi_{ab}$$

gives the coefficient of $\underline{L}$, and similarly for L. Then

$$2II = \langle D_a L, \chi_{bc} \underline{L} \rangle - \langle D_c L, \chi_{ab} \underline{L} \rangle.$$

Since

$$\langle D_a L, \underline{L} \rangle = \langle D_a T, T \rangle - \langle D_a T, N \rangle + \langle D_a N, T \rangle - \langle D_a N, N \rangle = 2k_{aN},$$

we find

$$II = k_{aN} \chi_{bc} - k_{cN} \chi_{ab}.$$

(b) In the formula

$$\not{D}_c \chi_{ab} - \not{D}_a \chi_{bc} = R_{bLca} - k_{cN} \chi_{ab} + k_{aN} \chi_{cb}$$

proved in (a), lift the index b and take $c = b$ to obtain

$$\not{D}_b \chi_a^b - \not{D}_a \chi_b^b = R_{bLba} - k_{bN} \chi_a^b + k_{aN} tr\, \chi .$$

We finally have to split χ in this formula. First,

$$e_a(\operatorname{tr}\chi) = e_a(\chi_b^b) = \not{D}_a \chi_b^b + 2\chi(e_b, \not{D}_a e_b).$$

Since $\langle \not{D}_a e_b, e_b \rangle = 0$ and $\langle \not{D}_1 e_1, e_2 \rangle = -\langle \not{D}_1 e_2, e_1 \rangle$, we find, for instance,

$$\chi(e_b, \not{D}_1 e_b) = \langle \not{D}_1 e_1, e_2 \rangle (\chi(e_1, e_2) - \chi(e_2, e_1)) = 0,$$

and similarly for $\not{D}_2$. Next,

$$\not{D}_b \chi = \not{D}_b \hat{\chi} + \tfrac{1}{2} e_b(\operatorname{tr}\chi) g,\ \not{D}_b \chi_a^b = \not{D}_b \hat{\chi}_a^b + \tfrac{1}{2} e_a(\operatorname{tr}\chi),$$

which yields the formula. ◇

2. Ellipticity We have to explain why the system on $\hat{\chi}$ is called "elliptic." Note that $\hat{\chi}$ depends only on two functions $\hat{\chi}_{11}$ and $\hat{\chi}_{12}$, and that we have two equations. More precisely, we know from the definition that

$$\not{D}_c \hat{\chi}_{ab} = e_c(\hat{\chi}_{ab}) + \cdots ,$$

where the dots stand for zero order terms in $\hat{\chi}$. Hence the Codazzi equations are

$$e_1(\hat{\chi}_{11}) + e_2(\hat{\chi}_{12}) + \cdots = \cdots ,\ e_1(\hat{\chi}_{12}) + e_2(\hat{\chi}_{22}) + \cdots = \ldots .$$

Since $\hat{\chi}$ is symmetric and traceless, this can be written as a first order 2×2 system on the unknowns $\hat{\chi}_{11}, \hat{\chi}_{12}$, with matrix

$$\begin{bmatrix} e_1 & e_2 \\ -e_2 & e_1 \end{bmatrix}.$$

The principal symbol of the matrix operator is the principal symbol of $e_1^2 + e_2^2$, that is, the principal symbol of the Laplace operator on the spheres.

3. The system on η

Theorem (div–curl system on η) *The 1-form η satisfies the following system:*

$$\operatorname{div}\eta = \tfrac{1}{2}\underline{L}(\operatorname{tr}\chi) - \tfrac{1}{2}k_{NN}\operatorname{tr}\chi - |\eta|^2 + \tfrac{1}{2}\hat{\chi}_{ab}\hat{\underline{\chi}}_{ab} + \tfrac{3}{4}(\operatorname{tr}\chi)(\operatorname{tr}\underline{\chi}) - \tfrac{1}{2}R^a{}_{L\underline{L}a},$$
$$\operatorname{curl}\eta_{ab} = \tfrac{1}{2}(\hat{\chi}_{bc}\hat{\underline{\chi}}_{ac} - \hat{\chi}_{ac}\hat{\underline{\chi}}_{bc}) + \tfrac{1}{2}(-R_{aL\underline{L}b} + R_{bL\underline{L}a}).$$

Proof (a) We first prove the formula

$$\not{D}_{\underline{L}}\chi_{ab} = 2\,\not{D}_a\eta_b + \chi_{ab}k_{NN} + 2\eta_a\eta_b - \underline{\chi}_{ac}\chi_{cb} + R_{bL\underline{L}a}.$$

Using the definitions,

$$\begin{aligned}\not{D}_{\underline{L}}\chi_{ab} &= \underline{L}(\chi_{ab}) - \chi(\not{D}_{\underline{L}}e_a, e_b) - \chi(e_a, \not{D}_{\underline{L}}e_b)\\ &= \langle D_{\underline{L}}D_a L, e_b\rangle + \langle D_a L, D_{\underline{L}}e_b\rangle - \chi(\not{D}_{\underline{L}}e_a, e_b) - \chi(e_a, \not{D}_{\underline{L}}e_b).\end{aligned}$$

Forcing the curvature term into the formula by adding and substracting terms, we get

$$\begin{aligned}\not{D}_{\underline{L}}\chi_{ab} &= R_{bL\underline{L}a} + \langle D_a D_{\underline{L}}L, e_b\rangle + \langle D_{[\underline{L},e_a]}L, e_b\rangle\\ &\quad + \langle D_a L, D_{\underline{L}}e_b\rangle - \chi(\not{D}_{\underline{L}}e_a, e_b) - \chi(e_a, \not{D}_{\underline{L}}e_b)\\ &= R_{bL\underline{L}a} + 2e_a(\eta_b) - \langle D_{\underline{L}}L, D_a e_b\rangle + \langle D_{[\underline{L},e_a]}L, e_b\rangle\\ &\quad + \langle D_a L, D_{\underline{L}}e_b\rangle - \chi(\not{D}_{\underline{L}}e_a, e_b) - \chi(e_a, \not{D}_{\underline{L}}e_b).\end{aligned}$$

The third, fourth, and fifth terms are handled by brute force, using the formula above for the frame coefficients:

$$\begin{aligned}\langle D_{\underline{L}}L, D_a e_b\rangle &= 2\eta(\not{D}_a e_b) - \chi_{ab}k_{NN},\\ [\underline{L}, e_a] &= \not{D}_{\underline{L}}e_a - \underline{\chi}_{ac}e_c + (\eta_a - k_{aN})(\underline{L} - L),\\ \langle D_a L, D_{\underline{L}}e_b\rangle &= \chi(e_a, \not{D}_{\underline{L}}e_b) + 2k_{aN}\eta_b.\end{aligned}$$

Substituting into the above formula, we see that the χ terms cancel out, while

$$e_a(\eta_b) - \eta(\not{D}_a e_b) = \not{D}_a\eta_b.$$

This yields the formula.

(b) We use the above formula in two ways. First, we take $b = a$ and sum to obtain

$$\not{D}_{\underline{L}}\chi_a^a = 2\mathrm{div}\,\eta + k_{NN}\mathrm{tr}\,\chi + 2|\eta|^2 - \chi_{ab}\underline{\chi}_{ab} + R^a_{\;L\underline{L}a}.$$

We now split χ and $\underline{\chi}$ as explained at the end of section 7.2.2:

$$\chi = \hat{\chi} + \tfrac{1}{2}(\mathrm{tr}\chi)g,\ \underline{\chi} = \underline{\hat{\chi}} + \tfrac{1}{2}(\mathrm{tr}\,\underline{\chi})g.$$

Noting that $\not{D}_{\underline{L}}\chi_a^a = \underline{L}(\mathrm{tr}\,\chi)$, we obtain the first formula of the theorem.

For the second, we substract the formula with ba from the formula with ab we have established, to get

$$2\mathrm{curl}\,\eta_{ab} = \chi_{bc}\underline{\chi}_{ac} - \chi_{ac}\underline{\chi}_{bc} - R_{aL\underline{L}b} + R_{bL\underline{L}a}.$$

Splitting χ and $\underline{\chi}$ as usual, the χ, $\underline{\chi}$ terms yield the same terms with χ replaced by $\hat{\chi}$ and $\underline{\chi}$ by $\underline{\hat{\chi}}$, the other terms cancelling out by symmetry. ◇

Note that the system on η is also elliptic, since its matrix operator has the same principal symbol as the Laplace operator on the spheres.

7.5 Mixed transport–elliptic systems

If we examine the system on $\mathrm{tr}\,\chi$, $\hat{\chi}$, η, we observe the presence of the term $\underline{L}(\mathrm{tr}\,\chi)$ in the expression for $\mathrm{div}\,\eta$, a term for which we have no control. Since we already know $L(\mathrm{tr}\,\chi)$, we can compute $L(\mathrm{div}\,\eta)$ and write $L(\underline{L}\mathrm{tr}\,\chi) = [L, \underline{L}]\mathrm{tr}\,\chi + \underline{L}(L\mathrm{tr}\,\chi)$. The result is summarized in the following theorem.

Theorem *Set $\mu_1 = \underline{L}(\mathrm{tr}\,\chi) - \frac{1}{2}(\mathrm{tr}\,\chi)^2$. The quantity μ_1 satisfies the following transport equation:*

$$\begin{aligned} L\mu_1 + (\mathrm{tr}\chi)\mu_1 = &-\underline{L}(R_{LL}) - 2\, \not{D}_{\underline{L}}\hat{\chi}_{ab}\hat{\chi}^{ab} + 2(\underline{\eta}_a - \eta_a)e_a(\mathrm{tr}\,\chi) \\ &- \underline{L}(k_{NN})(\mathrm{tr}\,\chi) - (k_{NN} + \mathrm{tr}\,\chi)L(\mathrm{tr}\,\chi) - \tfrac{1}{2}(\mathrm{tr}\,\chi)^3, \end{aligned}$$

where the quantity $\not{D}_{\underline{L}}\hat{\chi}$ is given by

$$\begin{aligned} \not{D}_{\underline{L}}\hat{\chi}_{ab} = &\,2\, \not{D}_a\eta_b - div\,\eta\delta_{ab} + k_{NN}\hat{\chi}_{ab} + 2(\eta_a\eta_b - |\eta|^2\delta_{ab}) \\ &- \tfrac{1}{2}(\mathrm{tr}\,\underline{\chi})\hat{\chi}_{ab} - \tfrac{1}{2}(\mathrm{tr}\,\chi)\underline{\hat{\chi}}_{ab} + R_{aL\underline{L}b}. \end{aligned}$$

Proof The formula for $\not{D}_{\underline{L}}\hat{\chi}$ follows from the formula in the proof of the theorem about η just by splitting χ:

$$\not{D}_{\underline{L}}\chi = \not{D}_{\underline{L}}\hat{\chi} + \tfrac{1}{2}(\underline{L}(\mathrm{tr}\,\chi)g + (\mathrm{tr}\,\chi)\, \not{D}_{\underline{L}}g).$$

Proving as usual $\not{D}_{\underline{L}}g = 0$, and using the expression for $\underline{L}(\mathrm{tr}\,\chi)$ in terms of $\mathrm{div}\,\eta$ given in the theorem, we obtain the formula.

To prove the transport formula, we check first

$$L\mu_1 + (\mathrm{tr}\,\chi)\mu_1 = L\underline{L}(\mathrm{tr}\,\chi) + \underline{L}(\mathrm{tr}\,\chi)(\mathrm{tr}\,\chi) - L(\mathrm{tr}\,\chi)(\mathrm{tr}\,\chi) - \tfrac{1}{2}(\mathrm{tr}\,\chi)^3.$$

On the other hand, using the formula for $L(\mathrm{tr}\,\chi)$, we get

$$\begin{aligned} \underline{L}(L(\mathrm{tr}\,\chi)) = &-\underline{L}(R_{LL}) - 2\, \not{D}_{\underline{L}}\hat{\chi}_{ab}\hat{\chi}^{ab} \\ &- (\mathrm{tr}\,\chi)[\underline{L}(\mathrm{tr}\,\chi) + \underline{L}(k_{NN})] - k_{NN}\underline{L}(\mathrm{tr}\,\chi). \end{aligned}$$

Using the formula

$$[L, \underline{L}] = 2(\underline{\eta}_a - \eta_a)e_a + k_{NN}(\underline{L} - L),$$

we finish the proof. ◇

The way one uses these formulas to obtain an actual control of all frame coefficients is far from obvious:

(i) We note the presence of terms $e_a(\mathrm{tr}\,\chi)$ in the transport equation for μ_1. Hence one has also to establish a transport equation for $\not{D}(\mathrm{tr}\,\chi)$. The system of the transport equations on $\operatorname{tr}\chi$, $e_a(\operatorname{tr}\chi)$, and the elliptic Codazzi equations on $\hat{\chi}$ and $\not{D}\hat{\chi}$ is closed.

(ii) To control η, we use the transport equation on μ_1 along with the elliptic system for η, which form a closed system on η, μ_1, and $\not{D}\eta$.

We refer the reader to [31] for the actual implementation of this strategy.

8

Nonlinear equations, a priori estimates and induction

In the preceding chapters, we considered the wave equation associated with a given Lorentzian metric g, and explained various ways of analyzing the qualitative behavior of the solutions of this equation. This *linear problem* is of interest in itself, and is far from being solved with some generality (see, for instance, [6] for an answer to a natural decay question). However, in the literature much emphasis has been put on *nonlinear problems*, that is on problems where the metric g depends on the solution itself. One of the most interesting of these problems is the Cauchy problem for the Einstein equations, where the metric g itself is the unknown. For these nonlinear problems, since the metric is not given a priori, we have to explain how the preceding techniques can be used. It turns out that the problem of the long time existence of solutions can be reduced to some a priori estimates of the solutions of a wave equation associated with a given metric. The main concept which makes this possible is that of "induction on time." We first explain this concept with a very simple ordinary differential equation (ODE) example; after that, to illustrate how the method works for PDE, we present a classical result on the lifespan of solutions to some quasilinear wave equation with small smooth data.

8.1 A simple ODE example

Consider the Cauchy problem for the system of ODEs

$$X'(t) = AX(t) + F(X(t)),\ X(0) = X_0,\ t \geq 0.$$

We assume here that A is an $N \times N$ real constant matrix, X is the unknown function on some (unknown) interval of $\mathbf{R}_t^+$, taking values in $\mathbf{R}^N$, and

$$F : \mathbf{R}^N \to \mathbf{R}^N$$

is assumed to be C^1.

We consider the case of an attractive equilibrium point, that is:

(i) all eigenvalues of A have strictly negative real parts;
(ii) $F(0) = 0$ and $|F(X)| \leq F_0 |X|^2$.

We can then prove the following theorem.

Theorem *There are constants $A_1 > 0$, $a > 0$ such that, if $|X_0|$ is small enough, the solution X exists for all $t \geq 0$ and satisfies*

$$|X(t)| \leq A_1 |X_0| e^{-at}.$$

Proof (a) First, we replace the Cauchy problem by some integral equation. Setting $X(t) = e^{At} Y(t)$, we get $X'(t) = Ae^{At} Y(t) + e^{At} Y'(t)$, hence

$$Y'(t) = e^{-At} F(X(t)),\ Y(0) = X_0.$$

Integrating this between 0 and t, and replacing Y, we obtain

$$X(t) = e^{At} X_0 + \int_0^t e^{A(t-s)} F(X(s)) ds.$$

If $X \in C^0(I)$ satisfies this integral equation, then, in fact, $X \in C^1(I)$ and X satisfies the original Cauchy problem.

(b) Next, we note that for some $a > 0$ and some constant A_0, $||e^{At}|| \leq A_0 e^{-at}$. This is a consequence of the linear algebra fact: there are an invertible matrix P, a nilpotent matrix N and a diagonal matrix D with

$$P^{-1} A P = D + N,\ [D, N] = DN - ND = 0.$$

Then, since D and N commute,

$$e^{At} = P e^{Dt} e^{Nt} P^{-1},$$

and e^{Nt} is a polynomial in t, while the diagonal elements of D, which are the eigenvalues of A, have all real parts less than, say, $-b < 0$. Since $||e^{Dt}|| \leq e^{-bt}$, we can take for a any positive number strictly smaller than b.

(c) We make now the *induction hypothesis*: For some C_0 to be chosen later, and some $T > 0$,

$$|X(t)| \leq C_0 e^{-at},\ 0 \leq t < T.$$

Note that, if $C_0 > |X_0|$, this hypothesis is satisfied for some $T > 0$. We show now that, for a properly chosen C_0, this hypothesis implies in fact

$$|X(t)| \leq \tfrac{1}{2} C_0 e^{-at},\ 0 \leq t \leq T.$$

Using the integral formulation of (a), we have

$$\begin{aligned}|X(t)| &\le A_0 e^{-at}|X_0| + A_0 F_0 \int_0^t e^{-a(t-s)} C_0^2 e^{-2as}\,ds \\ &\le \left[A_0|X_0| + A_0 F_0 \frac{C_0^2}{a}\right] e^{-at}.\end{aligned}$$

If we choose (independently of T)

$$C_0 = \frac{a}{4A_0F_0}, \; |X_0| \le \frac{a}{16A_0^2F_0}, \; |X_0| < C_0,$$

we obtain $|X(t)| \le \frac{1}{2}C_0 e^{-at}$.

(d) We finish now the proof of the theorem. C_0 is fixed, and let us fix $|X_0|$ satisfying the above inequalities. Consider the supremum $\bar{T}$ of T such that the induction hypothesis is satisfied for $0 \le t < T$. If $\bar{T} < \infty$, there is a contradiction, since the estimate from (c) shows that X extends to some interval $[0, \bar{T} + \epsilon]$ with $\epsilon > 0$ and $|X(t)| \le C_0 e^{-at}$ there. Hence $\bar{T} = \infty$.

Finally, using this estimate and writing $|F(X(t))| \le F_0 C_0 e^{-at}|X(t)|$, we obtain

$$|X(t)| \le A_0 e^{-at}|X_0| + C_0 F_0 e^{-at} \int_0^t |X(s)| ds,$$

which gives, setting $\phi(t) = e^{at}|X(t)|$,

$$\phi(t) \le A_0|X_0| + \int_0^t e^{-as}\phi(s) ds.$$

Using Gronwall's lemma, we get $\phi(t) \le A_0|X_0| \exp F_0(C_0/a)$, which yields the result with $A_1 = A_0 \exp(1/4A_0)$. ◇

The preceding proof uses two fundamental facts from the theory of ODE:

(i) The solution to the Cauchy problem exists locally.
(ii) If $|X(t)| \le M$ for some $t < T$, the solution X can be extended to some bigger interval $t \le T + \epsilon$.

Similarly, the process of "induction on time" for a PDE requires

(i) a local existence theory,
(ii) a blowup criterion.

We are going to study these two aspects separately.

8.2 Local existence theory

1. The standard theorem is stated in the context of the Cauchy problem for first order quasilinear symmetric systems

$$S(U)\partial_t U + \sum A_i(U)\partial_i U + B(U) = 0,\ U(x,0) = U_0(x).$$

Here, $U = U(x,t)$ takes values in $\mathbf{R}^N$, S and A_i are real symmetric $N \times N$ matrices depending smoothly on $U \in \mathbf{R}^N$, while B maps smoothly $\mathbf{R}^N$ to itself. Moreover, S is positive definite (such a system is called hyperbolic symmetric, see [9]).

Theorem *If $U_0 \in H^s(\mathbf{R}^n)$ for some $s > (n/2) + 1$, there exists $T > 0$ and a unique solution $U \in C^0([0,T], H^s) \cap C^1([0,T], H^{s-1})$ of the Cauchy problem. Moreover, T can be chosen to depend on $||U_0||_{H^s}$ only.*

Here, H^s denotes the Sobolev space. We refer the reader to [42] or to [10] for a proof.

2. We consider now two different types of nonlinear wave equations on $\mathbf{R}^4_{x,t}$:

Type (a) Equations of the form

$$\Box u + \sum g^{\alpha\beta\gamma}(\partial_\gamma u)(\partial^2_{\alpha\beta} u) = 0,$$

where the $g^{\alpha\beta\gamma}$ are real constants. Such equations will be hyperbolic if ∇u is small enough.

Type (b) Equations of the form

$$\Box u + \sum g^{\alpha\beta}(u)\partial^2_{\alpha\beta} u = 0,$$

where the $g^{\alpha\beta}$ are real C^∞ functions with $g^{\alpha\beta}(0) = 0$. Such equations will be hyperbolic if u is small enough.

a. A nonlinear wave equation of type (a) can be reduced to a first order hyperbolic symmetric system by introducing the new unknowns $U = (\partial_t u, \partial_1 u, \partial_2 u, \partial_3 u)$. The local existence result from the preceding section then translates into the following theorem.

Theorem *For a nonlinear wave equation of type (a), if, for some $s > 7/2$,*

$$u_0(x) = u(x,0) \in H^s,\ u_1(x) = (\partial_t u)(x,0) \in H^{s-1},$$

there exists $T > 0$ and a unique solution u to the Cauchy problem with data (u_0, u_1) satisfying

$$u \in C^0([0,T], H^s) \cap C^1([0,T], H^{s-1}).$$

b. A nonlinear wave equation of type (b) can be reduced to a first order hyperbolic symmetric system by introducing the new unknowns $U = (u, \partial_t u, \partial_1 u, \partial_2 u, \partial_3 u)$. However, this is not the best way to obtain a low regularity result, since u and ∇u are considered at the same level. To obtain the correct result, one has to imitate the proof of the theorem for the first order systems. We then obtain the following theorem.

Theorem *For a nonlinear equation of type (b), if, for some $s > 5/2$,*

$$u_0(x) = u(x, 0) \in H^s,\ u_1(x) = (\partial_t u)(x, 0) \in H^{s-1},$$

there exists $T > 0$ and a unique solution to the Cauchy problem with data (u_0, u_1) satisfying

$$u \in C^0([0, T], H^s) \cap C^1([0, T], H^{s-1}).$$

8.3 Blowup criteria

The question is the following:

> Suppose a solution u of some quasilinear wave equation or system exists and is smooth for $t < T$; what minimal condition on u would ensure that u can be extended smoothly beyond $t = T$?

This minimal condition is, in fact, a "nonblowup" criterion; the blowup criterion is obtained by negation: Assuming that T is the lifespan of the smooth solution u (that is, the maximal time during which u will be smooth), then the minimal condition is not satisfied. For instance, for an ODE, if $|X|$ does not go to $+\infty$ as $t \to T$, then u can be extended beyond T.

We now state blowup criteria for hyperbolic symmetric systems or nonlinear wave equations considered in the previous section.

1. Hyperbolic symmetric systems

Theorem *Consider a hyperbolic symmetric first order system as in section 8.2. Suppose that, for some $s > (n/2) + 1$, there exists a solution*

$$U \in C^0([0, T[, H^s) \cap C^1([0, T[, H^{s-1}).$$

Assume that there exists M such that, for $t < T$

$$|U(x, t)| \leq M, \quad \int^T ||\nabla_x U(\cdot, t)||_{L^\infty} dt < \infty.$$

Then, for some $\epsilon > 0$, U can be extended for $t \leq T + \epsilon$ with the same regularity.

Proof We give here a sketch of the proof, since it helps when trying to understand other cases as well with similar proofs. The idea is to control the

H^s norm of $U(\cdot, t)$ uniformly for $t < T$. If this can be done, the local existence theorem, applied with data given on the initial surface $\{t = t_0\}$ close enough to $\{t = T\}$, will yield the desired extension of U.

Step 1 We need to establish an energy inequality for the linearized equation

$$S(U)\partial_t V + \sum A_i(U)\partial_i V = F,\ V(x,0) = V_0(x).$$

To do this, we write first the differential identity

$${}^tVF = \tfrac{1}{2}\partial_t({}^tVSV) + \sum \tfrac{1}{2}\partial_i({}^tVA_iV) - \tfrac{1}{2}{}^tVGV,\ G = \partial_t S + \sum \partial_i A_i.$$

Integrating this identity in the strip between 0 and t, we get

$$E(t) - E(0) = \int_{0\le s\le t} {}^tVF\,dxds + \tfrac{1}{2}\int_{0\le s\le t} {}^tVGV\,dxds,$$

where the energy E is defined by

$$E(t) = \tfrac{1}{2}\int ({}^tVSV)(x,t)dx.$$

Since U is bounded and S positive definite, there exists a constant $\alpha > 0$ such that

$$\alpha||V(\cdot,t)||_{L^2}^2 \le E(t) \le \alpha^{-1}||V(\cdot,t)||_{L^2}^2.$$

Also, using the equation on U,

$$G = S'(U)\partial_t U + \sum A_i'(U)\partial_i U,\ ||G|| \le C(1 + |\nabla_x U|).$$

Using Gronwall's lemma, we finally obtain as usual, the energy inequality

$$||V(\cdot,t)||_{L^2} \le C\{||V_0||_{L^2} + \int_0^t ||F(\cdot,s)||_{L^2}ds\}\exp C\int_0^t ||\nabla_x U||_{L^\infty}ds.$$

We refer the reader to [9] for details.

Step 2 For simplicity, we assume $s \in \mathbf{N}$, and commute an operator ∂_x^α, $|\alpha| \le s$ with the equation for U. We obtain

$$S(U)\partial_t V + \sum A_i(U)\partial_i V = F$$

with $V = \partial_x^\alpha U$ and

$$F = -\sum_{|\beta|\ge 1} C_\alpha^\beta S\partial_x^\beta(S^{-1}A_i)\partial_i\partial_x^{\alpha-\beta}U - S\partial_x^\alpha(S^{-1}B).$$

At this stage, we need two lemmas about the derivatives of some nonlinear expressions.

Lemma 1 *For functions $u, v \in L^\infty \cap H^s$, and mutiindices α, β with $|\alpha| + |\beta| = s$, we have*

$$||(\partial_x^\alpha u)(\partial_x^\beta v)||_{L^2} \leq C\{||u||_{L^\infty}||v||_{H^s} + ||v||_{L^\infty}||u||_{H^s}\}.$$

Lemma 2 *For any C^∞ function F on $\mathbf{R}^N$ with $F(0) = 0$, and $u \in L^\infty \cap H^s$, we have*

$$||F(u)||_{H^s} \leq C_0||u||_{H^s},$$

where the constant C_0 depends only on F and $||u||_{L^\infty}$.

The proofs of these lemmas can be found in, for instance, [10], [42]. For the first term in F, assuming, for example, $\partial_x^\beta = \partial_x^\gamma \partial_j$, we write

$$\sum C_\alpha^\beta S\partial_x^\gamma(\partial_j(S^{-1}A_i))\partial_x^{\alpha-\beta}(\partial_i U).$$

Using lemma 1 for the index $s - 1$, we obtain the bound

$$||F(\cdot, t)||_{L^2} \leq C + C(1 + ||\nabla_x U(\cdot, t)||_{L^\infty})||U(\cdot, t)||_{H^s}.$$

Step 3 Using the energy inequality of step 1 to estimate $V = \partial_x^\alpha U$, and summing over all $|\alpha| \leq s$, we finally get

$$||U(\cdot, t)||_{H^s} \leq C\{C + ||U_0||_{H^s} + \int_0^t ||\nabla_x U(\cdot, t)||_{L^\infty}||U(\cdot, s)||_{H^s} ds\}.$$

Using Gronwall's lemma once more, this gives a fixed bound on $||U(\cdot, t)||_{H^s}$ for $t < T$ as desired. ◇

2. Nonlinear wave equations By translating the blowup criterion for hyperbolic symmetric systems, one obtains the following criterion for nonlinear wave equations of type (a).

Theorem *A solution u of a wave equation of type (a), supposed to be smooth for $t < T$, can be extended smoothly beyond T if*

$$|\nabla u| \leq M, \quad \int^T ||(\nabla^2 u)(\cdot, t)||_{L^\infty} < \infty.$$

For wave equations of type (b), a similar proof to the one given above for systems yields the following criterion.

Theorem *A solution u of a wave equation of type (b), supposed to be smooth for $t < T$, can be extended smoothly beyond T if*

$$|u| \leq M, \quad \int^T ||(\nabla u)(\cdot, t)||_{L^\infty} < \infty.$$

8.4 Induction on time for PDEs

For simplicity, we discuss here only C^∞ solutions. Consider the solution of a nonlinear wave equation with, say, C_0^∞ Cauchy data. A smooth solution u exists locally in time. The method of induction on time requires the definition, for $T > 0$, of a "minimal" property $\mathcal{P}_T$ of smooth solutions supposed to exist for $t < T$, in such a way that:

(i) $\mathcal{P}_T$ implies that u can be extended smoothly to $t \leq T + \epsilon$ for some $\epsilon > 0$;
(ii) if $\mathcal{P}_T$ is verified, one can, in fact, prove a stronger statement implying $\mathcal{P}_{T+\epsilon}$ for some $\epsilon > 0$.

The result is that u exists for all time and satisfies $\mathcal{P}_T$ for all T.

Let us consider the example of a nonlinear equation of type (a):

$$\Box u + \sum_{1 \leq i,j,k \leq 3} g^{ijk}(\partial_k u)(\partial_{ij}^2 u) = 0.$$

We will prove the following large time existence result due to Klainerman [25] (see also [24] for a more precise version).

Theorem *Let $u_0, u_1 \in C_0^\infty(\mathbf{R}^3)$ and consider the Cauchy problem with small data*

$$\Box u + \sum g^{ijk}(\partial_k u)(\partial_{ij}^2 u) = 0, \; u(x,0) = \epsilon u_0(x), \; (\partial_t u)(x,0) = \epsilon u_1(x).$$

There are constants $\epsilon_0 > 0$ and $C_0 > 0$ such that, for $\epsilon \leq \epsilon_0$, the solution u exists and is C^∞ for $t \leq \exp(C_0/\epsilon)$.

The essential point of the proof is to pick the correct induction hypothesis $\mathcal{P}_T$. Suppose, for some constant C_0 and s big enough, we require $||u(\cdot,t)||_{H^s} \leq C_0$ for $t < T$. Taking into account the above theorems, this indeed guarantees that u can be extended smoothly beyond T. To prove that the extension also satisfies $||u(\cdot,t)||_{H^s} \leq C_0$ with the same C_0 for $t < T + \epsilon$, all we can do is use energy inequalities to control the H^s norm of u. As we saw in the proof of the blowup criteria above, such an inequality contains the *amplification factor* $\exp C \int_0^T ||(\nabla^2 u)(\cdot,t)||_{L^\infty} dt$. Hence the only possibility of getting a reasonable bound from this energy inequality is to control this exponential factor, that is, to obtain a decay information (if possible, integrability) on $||(\nabla^2 u)(\cdot,t)||_{L^\infty}$. This can be done using the Klainerman inequality

$$(1+t+r)\langle r-t\rangle^{\frac{1}{2}}|v| \leq \sum ||Z^k v(\cdot,t)||_{L^2}$$

for $v = \nabla^2 u$. To obtain the required L^2 estimates on $\nabla Z^k u$, we have to commute products Z^k with the equation; this, in turn, produces nonlinear terms of the form $(\partial Z^p u)(\partial Z^q u)$. Control of these terms requires knowledge of the decay of $Z^k u$. Thus our induction hypothesis on the H^s norm of u will not work. We have to make an induction hypothesis *involving the Lorentz fields Z*.

Proof (a) The following commutation lemma describes how products Z^k of Lorentz fields commute with the equation.

Commutation lemma *There are constant coefficients $\tilde{g}$ such that, for all $k \geq 1$, the solution u satisfies the equation*

$$PZ^k u + \sum \tilde{g}(\partial Z^p u)(\partial^2 Z^q u) = 0, \ p \leq k-1, \ q \leq k-1, \ p+q \leq k.$$

Here

$$P = \Box + \sum g^{ijk}(\partial_k u)\partial^2_{ij} + \sum g^{ijk}(\partial^2_{ij} u)\partial_k$$

is the linearized operator on u, and the sum denotes symbolically sums of terms of the form $\tilde{g}^{\alpha\beta\gamma}_{pq}(\partial_\gamma Z^p u)(\partial^2_{\alpha\beta} Z^q u)$, the coefficients $\tilde{g}$ being, of course, different for each such term.

We prove the lemma by induction on k. Denoting by c various real constants, we note that for all Z fields,

$$[\Box, Z] = c\Box, \ [Z, \partial] = c\partial, \ [Z, \partial^2] = \sum c\partial^2.$$

To prove the last property we write

$$Z\partial^2_{\alpha\beta} = [Z, \partial_\alpha]\partial_\beta + \partial_\alpha[Z, \partial_\beta] + \partial^2_{\alpha\beta} Z,$$

and repeatedly use $[Z, \partial] = c\partial$. Thus, applying Z to the equation for u, we get

$$PZu + [Z, \Box]u + \sum g^{ijk}([Z, \partial_k]u)(\partial^2_{ij} u) + \sum g^{ijk}(\partial_k u)[Z, \partial^2_{ij}]u = 0,$$

which is the result of the lemma for $k = 1$.

Assuming the result for k, we prove it for $k+1$, by applying Z to the equation for $Z^k u$. We obtain, omitting some indices for simplicity,

$$\begin{aligned}
&PZ^{k+1}u + [Z, \Box]Z^k u + \sum g([Z, \partial]u)(\partial^2 Z^k u) + \sum g(\partial Zu)(\partial^2 Z^k u) \\
&\quad + \sum g(\partial u)([Z, \partial^2]Z^k u) + \sum g([Z, \partial]Z^k u)(\partial^2 u) + \sum g(\partial Z^k u)([Z, \partial^2]u) \\
&\quad + \sum g(\partial Z^k u)(\partial^2 Zu) + \sum \tilde{g}([Z, \partial]Z^p u)(\partial^2 Z^q u) + \sum \tilde{g}(\partial Z^{p+1} u)(\partial^2 Z^q u) \\
&\quad + \sum \tilde{g}(\partial Z^p u)([Z, \partial^2]Z^q u) + \sum \tilde{g}(\partial Z^p u)(\partial^2 Z^{q+1} u) = 0.
\end{aligned}$$

(b) We need an energy inequality for P. This is obtained similarly to what was done for first order systems. Multiplying Pv by $\partial_t v$, we obtain a differential identity

$$(Pv)(\partial_t v) = \tfrac{1}{2}\partial_t[(\partial_t v)^2 + \sum(\partial_i v)^2 - \sum g^{ijk}(\partial_k u)(\partial_i v)(\partial_j v)] \\ + \sum \partial_i[\cdots] + q.$$

Here, q is a quadratic form in ∇v with coefficients that are linear combinations (with constant coefficients) of second order derivatives of u. Integrating $(Pv)(\partial_t v)$ in the strip between 0 and T, we obtain

$$E(T) - E(0) = \int_{0\le t\le T} (Pv)(\partial_t v) dx dt - \int_{0\le t\le T} q dx dt,$$

where the natural energy E at time T is defined by

$$E(T) = \tfrac{1}{2}\int \{(\partial_t v)^2 + \sum(\partial_i v)^2 - \sum g^{ijk}(\partial_k u)(\partial_i v)(\partial_j v)\}(x, T) dx.$$

Hence we have to be careful about the smallness of ∇u: there exists a constant M_0 such that, if $|\nabla u| \le M_0$,

$$\alpha||(\nabla v)(\cdot, T)||_{L^2}^2 \le E(T) \le \alpha^{-1}||(\nabla v)(\cdot, T)||_{L^2}^2$$

for some fixed $\alpha > 0$. Assuming from now on that $|\nabla u| \le M_0$, we obtain as usual the energy inequality

$$||\nabla v(\cdot, t)||_{L^2} \le C\{||\nabla v(\cdot, 0)||_{L^2} \\ + \int_0^t ||(Pv)(\cdot, s)||_{L^2} ds\} \exp C \int_0^t ||(\nabla^2 u)(\cdot, s)||_{L^\infty} ds.$$

(c) We can now formulate our induction hypothesis involving the Lorentz fields. For some constant C_1 to be chosen later, and some big enough even integer N, assume

$$\mathcal{P}_T : \{\phi(t) \equiv \sum_{k\le N} ||(\partial Z^k u)(\cdot, t)||_{L^2} \le C_1\epsilon,\ t < T\}.$$

We take $k \le N$, and consider the equation $PZ^k u = F_k$, where F_k is given by the commutation lemma.

Lemma *If* $N \ge 6$,

$$||F_k(\cdot, t)||_{L^2} \le CC_1\epsilon(1+t)^{-1} \sum_{l\le k} ||(\partial Z^l u)(\cdot, t)||_{L^2}.$$

First, using the Klainerman inequality, we note that for $l \le N - 2$,

$$(1+t)|\partial Z^l u| \le C \sum_{m\le 2} ||Z^m \partial Z^l u||_{L^2} \le CC_1\epsilon.$$

Consider one term $\tilde{g}(\partial Z^p u)(\partial^2 Z^q u)$ in F_k; since $p + q \leq k \leq N$, either $p \leq N/2$ or $q \leq N/2$. In the first case, we can use the L^∞ bound on $\partial Z^p u$ if $N/2 \leq N - 2$, that is, $N \geq 4$; in the second case, we can use the L^∞ bound on $\partial^2 Z^q u$ if $N/2 + 1 \leq N - 2$, that is $N \geq 6$. In all cases, we obtain the conclusion of the lemma.

(d) To use the energy inequality for the equation $PZ^k u = F_k$, we need to estimate $\nabla Z^k u$ at time $t = 0$. To do this, it is enough to remark that the equation on u allows one to express $\partial_t^2 u$ in terms of the data u_0, u_1, and, more generally, all t-derivatives. Hence, for some constant C_2,

$$\sum_{k \leq N} ||(\partial Z^k u)(\cdot, 0)||_{L^2} \leq C_2 \epsilon.$$

Note that the induction hypothesis implies, in particular, $|\partial u| \leq M_0$ for ϵ small enough, which we assume from now on. We can then use the energy inequality for P to estimate $\partial Z^k u$ from the equation $PZ^k u = F_k$. Summing all the terms, we get

$$\phi(t) \leq \left\{ CC_2\epsilon + CC_1\epsilon \int_0^t \frac{\phi(s)}{1+s} ds \right\} \exp CC_1\epsilon \log(1+t).$$

Using Gronwall's lemma, we finally get, with numerical constants A, B,

$$\phi(t) \leq AC_2\epsilon \exp BC_1\epsilon \log(1+t).$$

We first choose $C_1 = 2AC_2$, and then ϵ small enough to make sure that $|\nabla u| \leq M_0$; finally, we take C_0 such that, for $t \leq T = \exp(C_0/\epsilon)$,

$$\exp BC_1\epsilon \log(1+t) \leq \tfrac{3}{2}.$$

These choices imply $\phi(t) \leq \frac{3}{4} C_1 \epsilon$. Hence the solution u can be extended as a smooth solution satisfying $\mathcal{P}_T$ for $T = \exp(C_0/\epsilon)$. ◇

Note that in this theorem, the solution is not global, but the process of induction on time works equally well until we reach $T = \exp(C_0/\epsilon)$. This limitation is not an artefact of the proof. If we denote by T_ϵ the maximal time for which the solution u exists and is smooth for $t < T_\epsilon$ (the "lifespan" of u), it can be shown ([1], [2]) under reasonable generic assumptions on the data u_0, u_1 that $\lim \epsilon \log T_\epsilon = l$ exists when $\epsilon \to 0$. In fact, the number l, conjectured by Hörmander [24], can be computed explicitly!

9
Applications to some quasilinear hyperbolic problems

As explained in the introduction, it is not possible here to give complete proofs of delicate results, some of which are several hundred pages long. We just want to point out how the methods and ideas explained in the preceding chapters enter the proofs of these results in an essential way. In the preceding chapter, we saw how the method of induction on time allows one to handle nonlinear problems as if they were linear problems. All problems discussed here will be handled in this framework. For each example below, we give a very brief sketch of the method of proof; we explain what is the null frame or the optical functions used in the work, how it is constructed, and why this frame is supposed to be a good frame. The examples we have chosen to discuss do not, of course, constitute the whole literature on the subject, but they seem representative. We have made no attempt to quote all works related to the ones we discuss, our purpose being only to illustrate; we hope to be forgiven for that. In the following list, to facilitate an overview, we characterize the method in a few words only.

Example 1 Global existence for small solutions of quasilinear wave equations

$$-\partial_t^2\phi + \Delta\phi + \sum g^{ij}(\partial\phi)\partial_{ij}^2\phi = 0$$

satisfying the null condition. The proof is by commuting *standard* Lorentz fields to get decay estimates.

Example 2 Global existence for small solutions of quasilinear wave equations

$$g^{\alpha\beta}(\phi)\partial_{\alpha\beta}^2\phi = 0.$$

Though the first proof used *modified* Lorentz fields, a new simpler proof uses only the *standard* Lorentz fields to get decay estimates.

Example 3 Low regularity well-posedness for quasilinear wave equations

$$-\partial_t^2\phi + \Delta\phi + \sum g^{ij}(\phi)\partial_{ij}^2\phi = N(\phi, \partial\phi).$$

The proof uses the full machinery of chapter 7 to obtain *decay* of solutions of some linear wave equation $\Box_{h_\lambda}$. Here, h_λ is a smoothed rescaled version of g.

Example 4 Stability of Minkowski space-time (first version). The full machinery of chapter 7 is used to prove decay for the solutions of the Bianchi equations, a first order system on the curvature tensor R.

Example 5 The L^2 conjecture for Einstein equations. The machinery of chapter 7 is used to control the geometry and the solutions of Bianchi equations in the context of very low regularity.

Example 6 Stability of Minkowski space-time (second version). Just as in examples 1 and 2, *standard* Lorentz fields are used to get decay estimates.

Example 7 Formation of black holes. The full machinery using the sphere foliation associated to two optical functions is used.

As shown by examples 1, 2, and 6, it turns out that, surprisingly enough, for some nonlinear *perturbation* problems, one can ignore the geometry of the linearized operator and work with the standard Lorentz fields. For example 2, and even more for example 6, this came as a surprise. This is due to the specific nonlinear structure of the equations, and to the fact that we are dealing with *small* solutions.

In constrast to these examples, examples 3, 4, 5, and 7 show that the geometric machinery explained in this book can be used in many different contexts:

(a) to prove *decay estimates* and global existence of solutions;
(b) to prove low regularity well-posedness, counting carefully derivatives;
(c) to prove the formation of singularities.

9.1 Quasilinear wave equations satisfying the null condition

The first pioneering works on this subject are due to Christodoulou [14] and Klainerman [26]. Consider the Cauchy problem with small data for a quasilinear wave equation

$$\partial_t^2\phi - \Delta\phi + \sum_{1\le i,j\le 3} g^{ij}(\partial\phi)\partial_{ij}^2\phi = 0,\ \phi(x,0) = \epsilon\phi_0(x),\ (\partial_t\phi)(x,0) = \epsilon\phi_1(x).$$

We assume $\phi_i \in C_0^\infty$ $(i = 1,2)$; for simplicity (cubic terms playing no role), we take

$$g^{ij}(\partial\phi) = \sum_{1\le k\le 3} g^{ijk}\partial_k\phi.$$

We restrict ourselves to nonlinear terms involving only tangential derivatives. This is only for simplicity, and has no special meaning.

1. The null condition Consider a set $g = (g^{\alpha\beta\gamma})$ $(0 \le \alpha, \beta, \gamma \le 3)$ of constant real coefficients.

Definition The *set* g is said to satisfy the null condition if, for all $\xi \in \mathbf{R}^4$ with

$$\xi_0^2 = \xi_1^2 + \xi_2^2 + \xi_3^2,$$

we have $g^{\alpha\beta\gamma}\xi_\alpha\xi_\beta\xi_\gamma = 0$.

The equation is said to satisfy the null condition if, for all $\xi \in \mathbf{R}^3$,

$$g^{ijk}\xi_i\xi_j\xi_k = 0.$$

The null condition can be interpreted by saying that the function $u = r - t$ is closer to being an optical function than it would be in the general case of a quasilinear wave equation. In fact,

$$\begin{aligned}\langle \nabla u, \nabla u\rangle &= g^{ij}(\partial\phi)\omega_i\omega_j = g^{ijk}\omega_i\omega_j(\partial_k\phi) = g^{ijk}\omega_i\omega_j(\partial_k\phi - \omega_k\partial_r\phi)\\ &= O(t^{-1}|Z\phi|).\end{aligned}$$

This suggests that we can work with the standard null frame of the Minkowski metric, as will be seen later in the proof of the main result.

On the other hand, the null condition indicates cancellations in the nonlinear terms, when evaluated on a free solution. In fact, if $\phi = r^{-1}F(r - t, \omega, r^{-1})$ is a solution of $\Box\phi = 0$ (see [9], [24]), then (with $\sigma = r - t$)

$$g^{\alpha\beta\gamma}(\partial_\gamma\phi)(\partial^2_{\alpha\beta}\phi) = r^{-2}[g^{\alpha\beta\gamma}\omega_\alpha\omega_\beta\omega_\gamma](\partial_\sigma F)(\partial_\sigma^2 F) + O(r^{-3}) = O(r^{-3}).$$

We explain now two basic facts about the null condition.

a. *Null condition and Lorentz fields*

Estimation lemma *If g satisfies the null condition, we have for any two smooth functions ϕ, ψ,*

$$|g^{\alpha\beta\gamma}(\partial_\gamma\phi)(\partial^2_{\alpha\beta}\psi)| \le C(1+t)^{-1}\big(|Z\phi||\partial^2\psi| + |\partial\phi||Z\partial\psi|\big).$$

Proof (a) Set

$$T_\alpha = \partial_\alpha + \omega_\alpha\partial_t,\ \omega_i = \frac{x^i}{r},\ \omega_0 = -1,$$

so that $T_0 = 0$. We write

$$T_i = t^{-1}(H_i - x^i\partial_t) + \omega_i\partial_t = t^{-1}[H_i - \omega_i(r - t)\partial_t].$$

Modulo T_i, all spatial derivatives can be replaced by t derivatives:

$$\partial_i = T_i - \omega_i \partial_t, \ \partial_{ij}^2 = T_i \partial_j - \omega_i T_j \partial_t + \omega_i \omega_j \partial_t^2.$$

Hence, modulo T_i,

$$g^{\alpha\beta\gamma}(\partial_\gamma \phi)(\partial_{\alpha\beta}^2 \psi) \equiv -g^{\alpha\beta\gamma}\omega_\alpha\omega_\beta\omega_\gamma(\partial_t\phi)(\partial_t^2\psi) = 0,$$

since $\omega_0^2 = 1 = \sum \omega_i^2$.

(b) From the identities

$$(t+r)L = S + \sum \omega_i H_i, \ (t-r)\underline{L} = S - \sum \omega_i H_i, \ \frac{R}{r} = t^{-1}\omega \wedge H,$$

we obtain the inequality (see also [9])

$$|r-t||\nabla\phi| \le C \sum_{k\le 1} |Z^k \phi|.$$

In particular, we also have

$$|T\phi| \le C(1+t)^{-1} \sum_{k\le 1} |Z^k\phi|.$$

Repeatedly using this inequality to bound the terms

$$g^{\alpha\beta\gamma}(T_\gamma\phi)(\partial_{\alpha\beta}^2\psi), \ T_\alpha\partial_\beta\psi, \ \partial_t T_\beta \psi,$$

we obtain the result. ◇

b. *Null condition and commutators*

Commutation lemma *If g satisfies the null condition, we have for any Lorentz field Z*

$$Z[g^{\alpha\beta\gamma}(\partial_\gamma\phi)(\partial_{\alpha\beta}^2\psi)] = g^{\alpha\beta\gamma}(\partial_\gamma Z\phi)(\partial_{\alpha\beta}^2\psi) + g^{\alpha\beta\gamma}(\partial_\gamma\phi)(\partial_{\alpha\beta}^2 Z\psi) + \tilde{g}^{\alpha\beta\gamma}(\partial_\gamma\phi)(\partial_{\alpha\beta}^2\psi),$$

where the new sum with constant coefficients $\tilde{g} = (\tilde{g}^{\alpha\beta\gamma})$ satisfies again the null condition.

Proof To the coefficients $g^{\alpha\beta\gamma}$ we associate the function $p(\xi) = g^{\alpha\beta\gamma}\xi_\alpha\xi_\beta\xi_\gamma$. The null condition is satisfied if and only if p is identically zero on the cone $\xi_0^2 = \sum \xi_i^2$. On the other hand,

$$\tilde{g}^{\alpha\beta\gamma}(\partial_\gamma u)(\partial_{\alpha\beta}^2 v) = g^{\alpha\beta\gamma}([Z, \partial_\gamma]u)(\partial_{\alpha\beta}^2 v) + g^{\alpha\beta\gamma}(\partial_\gamma u)([Z, \partial_{\alpha\beta}^2]v).$$

Recall the expression of the Poisson bracket of two functions $f(x,\xi)$ and $g(x,\xi)$:

$$\{f, g\} = (\partial_\xi f)(\partial_x g) - (\partial_x f)(\partial_\xi g).$$

Let z denote the symbol of the field Z, that is, for instance, $z = x^\alpha \xi_\alpha$ for $Z = S$, or $z = x^0\xi_i + x^i\xi_0$ for $Z = H_i$, etc. Since $[Z, \partial_\alpha]$ has symbol $-\partial_\alpha z = \{z, \xi_\alpha\}$, we see that the symbol $\tilde{p}$ associated with the coefficients $\tilde{g}$ is just $\{z, p\}$. Since $p = 0$ on the cone $\xi_0^2 = \sum \xi_i^2$, $\partial_\xi p$ is colinear to $(\xi_1, \xi_2, \xi_3, -\xi_0)$. This implies that $\tilde{p}$ also vanishes on the cone. In fact, for $Z = S$, $\{z, p\} = \xi \cdot (\partial_\xi p) = 0$; for $Z = R_1$ for instance, $\{z, p\} = \xi_3(\partial_{\xi_2} p) - \xi_2(\partial_{\xi_3} p) = 0$; for $Z = H_1$ for instance, $\{z, p\} = \xi_1(\partial_{\xi_0} p) = \xi_0(\partial_{\xi_1} p) = 0$. ◇

2. Energy inequality Exactly as in the proof of the large time existence in section 8.4, in order to control terms like $Z^k\phi$, we need an energy inequality for the linearized operator

$$P = \Box + \sum g^{ijk}(\partial_k\phi)\partial^2_{ij} + \sum g^{ijk}(\partial^2_{ij}\phi)\partial_k.$$

Suppose now that ϕ behaves roughly like a free solution of $\Box$. The null condition implies then a remarkable energy inequality for P.

Lemma (energy inequality) *Suppose the null condition on (g^{ijk}) is satisfied. Assume, for some constant C_0,*

$$\sum_{k \le 3} ||(\partial Z^k\phi)(\cdot, t)||_{L^2} \le C_0\epsilon, \ t < T.$$

Then, for $\eta > 0$ small enough, the following energy inequality for P holds for $t < T$:

$$||(\nabla\psi)(\cdot, t)||_{L^2} + \left\{ \int_{0\le s\le t} \langle r - s\rangle^{-1-\eta} \sum (T_i\psi)^2 dxds \right\}^{\frac{1}{2}}$$
$$\le C_\eta \left(||\nabla\psi(\cdot, 0)||_{L^2} + \int_0^t ||(P\psi)(\cdot, s)||_{L^2} ds \right) e^{CC_0\epsilon}.$$

The remarkable fact here is that there is almost *no amplification factor* in this inequality, despite the variable coefficients of P (compare with the inequalities obtained in chapter 8). This inequality is very similar to that we proved in section 5.3. In fact, the proof employs the "ghost weight" technique of section 5.3, using the approximate optical function $u = t - r$. Instead of verifying the "good multiplier condition," we give here a direct and elementary proof of the inequality.

Proof (a) With $a = a(t - r)$ to be chosen, we first establish the differential identity

$$e^a(P\psi)(\partial_t\psi) = \tfrac{1}{2}\partial_t\{e^a[(\partial_t\psi)^2 + \sum(\partial_i\psi)^2 - \sum g^{ijk}(\partial_k\phi)(\partial_i\psi)(\partial_j\psi)]\}$$
$$+ \sum \partial_i[\cdots] + e^a q,$$

where $q = q_1 - (a'/2)q_2$, and

$$q_1 = -\sum g^{ijk}(\partial_{ik}^2\phi)(\partial_j\psi)(\partial_t\psi) + \sum g^{ijk}(\partial_{ij}^2\phi)(\partial_k\psi)(\partial_t\psi) + \tfrac{1}{2}\sum g^{ijk}(\partial_{tk}^2\phi)(\partial_i\psi)(\partial_j\psi),$$

$$q_2 = \sum(T_i\psi)^2 - \sum g^{ijk}(\partial_k\phi)(\partial_i\psi)(\partial_j\psi) - 2\sum g^{ijk}(\partial_k\phi)\omega_i(\partial_j\psi)(\partial_t\psi).$$

This is done as usual:

$$e^a(\partial_t^2\psi)(\partial_t\psi) = \tfrac{1}{2}\partial_t(e^a(\partial_t\psi)^2) - \tfrac{1}{2}a'e^a(\partial_t\psi)^2,$$
$$e^a(\partial_i^2\psi)(\partial_t\psi) = -\tfrac{1}{2}\partial_t(e^a(\partial_i\psi)^2) + \partial_i(\cdots) + e^a a'[\omega_i(\partial_i\psi)(\partial_t\psi) + \tfrac{1}{2}(\partial_i\psi)^2].$$

This gives first

$$e^a(\Box\psi)(\partial_t\psi) = \tfrac{1}{2}\partial_t[(\partial_t\psi)^2 + \sum(\partial_i\psi)^2] + \partial_i[\cdots] - \tfrac{1}{2}a'e^a\sum(T_i\psi)^2.$$

For the additional terms in P, we have similarly

$$e^a(\partial_k\phi)(\partial_{ij}^2\psi)(\partial_t\psi) = \partial_i(\cdots) - e^a(\partial_{ik}^2\phi)(\partial_j\psi)(\partial_t\psi) - e^a(\partial_k\phi)((\partial_j\psi)(\partial_{ti}^2\psi) + e^a a'\omega_i(\partial_k\phi)(\partial_j\psi)(\partial_t\psi).$$

Now

$$e^a(\partial_k\phi)(\partial_j\psi)(\partial_{ti}^2\psi) = \partial_t[e^a(\partial_k\phi)(\partial_i\psi)(\partial_j\psi)] - e^a(\partial_{tk}^2\phi)(\partial_i\psi)(\partial_j\psi) - e^a(\partial_k\phi)(\partial_i\psi)(\partial_{tj}^2\psi) - e^a a'(\partial_k\phi)(\partial_i\psi)(\partial_j\psi),$$

so that summing in i, j gives us twice the desired term.

(b) We now choose $a'(s) = -A\langle s\rangle^{-1-\eta}$ for some small $\eta > 0$ and some large A. We claim then the inequality

$$q \geq \alpha(t)|\nabla\psi|^2, \quad \int_0^{+\infty}|\alpha(t)|dt \leq CC_0\epsilon.$$

To prove this claim, we first handle the terms q_1 in q not containing a'. We have

$$q_1 = \sum g(\partial^2\phi)(T\psi)(\partial\psi) + (\partial_t\psi)^2\beta_1,$$
$$\beta_1 = \sum g^{ijk}\omega_j(\partial_{ik}^2\phi) - \sum g^{ijk}\omega_k(\partial_{ij}^2\phi) + \tfrac{1}{2}\sum g^{ijk}\omega_i\omega_j(\partial_{tk}^2\phi).$$

If we replace ∂_i by $-\omega_i\partial_t$ in the expression for β_1, the resulting expression is

$$(\partial_t^2\phi)[\sum g^{ijk}\omega_i\omega_j\omega_k - \sum g^{ijk}\omega_i\omega_j\omega_k - \tfrac{1}{2}\sum g^{ijk}\omega_i\omega_j\omega_k] = 0.$$

Hence $|\beta_1| \leq C|T\partial\phi|$. Using the assumption of the lemma and the Klainerman inequality, we get

$$\sum_{k\leq 1}|Z^k\partial\phi| \leq CC_0\epsilon(1+t)^{-1}.$$

Since we saw in the proof of the estimation lemma that $|T\partial\phi| \leq C(1+t)^{-1}|Z\partial\phi|$, we obtain $|\beta_1| \leq CC_0\epsilon(1+t)^{-2}$.

On the other hand, since

$$|\partial^2\phi| \leq C\langle r-t\rangle^{-1}\sum_{k\leq 1}|Z^k\partial\phi| \leq CC_0\epsilon\langle r-t\rangle^{-1}(1+t)^{-1},$$

we have

$$\begin{aligned}|g(\partial^2\phi)(T\psi)(\partial\psi)| &\leq C\langle r-t\rangle^{-1-\eta}(T\psi)^2 + C\langle r-t\rangle^{1+\eta}(\partial^2\phi)^2(\partial\psi)^2\\ &\leq C\langle r-t\rangle^{-1-\eta}(T\psi)^2 + CC_0^2\epsilon^2(1+t)^{-2}|\nabla\psi|^2.\end{aligned}$$

Summarizing, we have

$$q_1 \geq -C\langle r-t\rangle^{-1-\eta}|T\psi|^2 - CC_0\epsilon(1+t)^{-2}|\nabla\psi|^2.$$

We handle the terms in q_2 similarly:

$$\begin{aligned}q_2 &= \sum(T_i\psi)^2 + \sum g(\partial\phi)(T\psi)(\partial\psi) + (\partial_t\psi)^2\beta_2,\\ \beta_2 &= 2\sum g^{ijk}(\partial_k\phi)\omega_i\omega_j - \sum g^{ijk}(\partial_k\phi)\omega_i\omega_j.\end{aligned}$$

Since $|T\phi| \leq C(1+t)^{-1}|Z\phi| \leq CC_0\epsilon\langle r-t\rangle^{\frac{1}{2}}(1+t)^{-2}$, we get again

$$|a'||\beta_2| \leq CC_0\epsilon(1+t)^{-2}.$$

On the other hand, with arbitrarily small $\eta_1 > 0$,

$$|g(\partial\phi)(T\psi)(\partial\psi)| \leq \eta_1(T\psi)^2 + CC_0^2\epsilon^2(1+t)^{-2}(\partial\psi)^2.$$

Hence

$$-\frac{a'}{2}q_2 \geq -\frac{a'}{2}(1-\eta_1)|T\psi|^2 - CC_0\epsilon(1+t)^{-2}|\nabla\psi|^2.$$

(c) To finish the proof, we take A big enough, and use Gronwall's lemma which gives the amplification

$$\exp CC_0\epsilon\int_0^t\frac{ds}{(1+s)^2} \leq \exp CC_0\epsilon. \qquad \diamond$$

3. Global existence We have now the tools to prove the following global existence result due to Christodoulou [14] and Klainerman [26].

Theorem *Suppose the nonlinear wave equation satisfies the null condition. Then, for all N, there exist $C_N > 0$, $\epsilon_N >$ such that, for $\epsilon \leq \epsilon_N$, the Cauchy problem admits a unique global C^∞ solution ϕ satisfying*

$$\sum_{k\leq N}||(\nabla Z^k\phi)(\cdot,t)||_{L^2} \leq C_N\epsilon.$$

Proof We proceed by induction on time. Set for convenience

$$E_N(t) = \frac{1}{2} \sum_{k \leq N} \int |(\partial Z^k \phi)(x,t)|^2 dx.$$

Our induction hypothesis is:

Induction hypothesis $\mathcal{P}_T$ *Assume that the solution* ϕ *exists for* $0 \leq t < T$ *and satisfies there, for some big* C_0 *and* N *to be chosen later,*

$$E_N(t) \leq C_0^2 \epsilon^2.$$

Using the Klainerman inequality, the induction hypothesis implies

$$|Z^l \partial \phi|(x,t) \leq C C_0 \epsilon (1+t)^{-1} \langle r - t \rangle^{-\frac{1}{2}}, \ l \leq N-2.$$

The idea now is to commute with the equation products Z^l of standard Lorentz fields ($l \leq N$). Defining the linearized operator P by

$$P \equiv \Box + g^{ijk} \partial_k \phi \partial_{ij}^2 + g^{ijk} \partial_{ij}^2 \phi \partial_k$$

and repeatedly using the algebraic lemma above about the null condition and commutators, we obtain an equation

$$P Z^l \phi = \sum_{p+q \leq l-1} h_{pq}^{\alpha\beta\gamma} (\partial_\gamma Z^p \phi)(\partial_{\alpha\beta}^2 Z^q \phi) \equiv H,$$

where, for each (p,q) and all $\xi \in \mathbf{R}^4$ satisfying $\xi_0^2 = \sum \xi_i^2$,

$$\sum h_{pq}^{\alpha\beta\gamma} \xi_\alpha \xi_\beta \xi_\gamma = 0.$$

If $N \geq 3$, then $(l-1)/2 \leq N-2$, and we can use the induction hypothesis and the estimation lemma to bound the factor in H containing fewer Z-fields than the other, thus obtaining, for some integrable $h(t)$,

$$|H| \leq C \epsilon h(t) \sum_{p \leq l} |\partial Z^p \phi|, \quad \int_0^{+\infty} h(t) dt \leq C C_0.$$

Using the energy inequality for P, we finally obtain the estimate

$$E_N(t) \leq C_1 E_N(0) e^{C C_0 \epsilon},$$

where C_1 is independent of C_0.

To finish the proof, it is enough to choose first $C_0^2 = 2 C_1 E_N(0)/\epsilon^2$, then ϵ small enough to obtain

$$E_N(t) \leq \frac{2}{3} C_0^2 \epsilon^2.$$

This shows that the solution ϕ is global. ◇

The heuristic of this proof is this: the solution being small, of order ϵ, the nonlinear terms are negligible for a long time (in fact, a time of order $\exp(1/\epsilon)$), hence the solution essentially behaves like a free solution. Because of the null condition, the nonlinear terms evaluated on this free solution are too weak to modify the behavior of the solution at any later time. So the solution is global as if we were dealing with a linear equation. One should be careful, however, that this heuristic does not yield global solutions in space dimension 2, due to a weaker decay rate of the free solutions. One can consult [2] for a complete analysis of this case. Note, however, that the proof given here works equally well in space dimension 2, which is not the case with proofs based on conformal energy inequalities.

9.2 Quasilinear wave equations

We consider the Cauchy problem for the quasilinear wave equation

$$g^{\alpha\beta}(\phi)\partial^2_{\alpha\beta}\phi = 0,\ \phi(x,0) = \epsilon\phi_0(x),\ (\partial_t\phi)(x,0) = \epsilon\phi_1(x).$$

The coefficients $g^{\alpha\beta}(s)$ are given C^∞ functions of one real variable s, with

$$g^{\alpha\beta}(0) = m^{\alpha\beta},$$

m being here the Minkowsksi metric, that is, $m^{\alpha\beta}\partial^2_{\alpha\beta} = -\partial_t^2 + \Delta_x$. The data ϕ_0, ϕ_1 are fixed functions in C_0^∞. The general result is the following.

Theorem ([3], [39]) *There exists $\epsilon_0 > 0$ such that, if $\epsilon \leq \epsilon_0$, there is a global C^∞ smooth solution to the Cauchy problem.*

Note that it makes a big difference whether the coefficients g depend on ϕ or, as in the preceding example, on $\nabla\phi$; for instance, consider the model equation

$$-\partial_t^2\phi + c^2(\partial_t\phi)\Delta\phi = 0.$$

Taking the t-derivative and setting $\psi = \partial_t\phi$, we get

$$-\partial_t^2\psi + c^2(\psi)\Delta\psi = -2\frac{c'}{c}(\psi)(\partial_t\psi)^2.$$

In other words, the left-hand side of the equation is of the form considered in this section, but there is a *source term* in the right-hand side, which makes even the small solutions blowup in finite time (see [1]). Einstein equations in "harmonic coordinates" are a system of such equations, for which the source terms display some sort of a null condition. We will discuss this later in this chapter.

The theorem was first proved in [3] in the special case

$$-\partial_t^2\phi + c^2(\phi)\Delta\phi = 0,$$

using *modified Lorentz fields*. However, Lindblad [39] has given a simpler proof, the geometrical aspects of which we discuss now. The starting point of the proof of [39] is the following *bet*, which is far from being obvious:

One can use the standard Lorentz fields Z and commute them with the equation.

Assuming this is true, it means that the good derivatives of ϕ will be the standard ones $L\phi = \partial_t\phi + \partial_r\phi, (R/r)\phi$, as explained in section 5.1. This example shows that, for a given problem, it is not clear beforehand how to choose the geometry of the relevant fields.

The proof is by induction on time. Following [39], we formulate the induction hypothesis as

$$E_N(t) \equiv \tfrac{1}{2}\sum_{k\leq N}\int |(\partial Z^k\phi)(x,t)|^2 dx \leq 16N\epsilon^2(1+t)^\delta$$

for some $0 < \delta < 1$.

1. Decomposition of $\Box$ One proceeds to express the linear operator $\tilde{\Box} \equiv g^{\alpha\beta}\partial^2_{\alpha\beta}$ using the standard derivatives L, $\underline{L}$, R/r. One should be careful that we do not consider the metric $g_{\alpha\beta}$ (inverse matrix of $g^{\alpha\beta}$), but consider $g^{\alpha\beta}$ just as a symmetric 2-tensor on the background manifold $\mathbf{R}^4$ with the flat Minkowski metric m. In particular, we define

$$g_{\alpha\beta} = m_{\alpha\alpha'}m_{\beta\beta'}g^{\alpha'\beta'}.$$

To express $g^{\alpha\beta}$ and $\tilde{\Box}$ in terms of the fields L, $\underline{L}$, e_a, we compute the double trace

$$g^{\alpha\beta} = m^{\alpha\alpha'}m^{\beta\beta'}g_{\alpha'\beta'}$$

in the standard null frame $(e_1, e_2, \underline{L}, L)$. We thus get

$$\begin{aligned} g^{\alpha\beta} &= \tfrac{1}{4}g_{LL}m^\alpha_{\underline{L}}m^\beta_{\underline{L}} + \tfrac{1}{4}g_{\underline{L}\underline{L}}m^\alpha_L m^\beta_L + \tfrac{1}{4}g_{L\underline{L}}(m^\alpha_L m^\beta_{\underline{L}} + m^\alpha_{\underline{L}}m^\beta_L) \\ &\quad - \tfrac{1}{2}g_{aL}(m^\alpha_{\underline{L}}m^\beta_a + m^\alpha_a m^\beta_{\underline{L}}) - \tfrac{1}{2}g_{\underline{L}a}(m^\alpha_L m^\beta_a + m^\alpha_a m^\beta_L) + g_{ab}m^{\alpha a}m^{\beta b}. \end{aligned}$$

Since, for any field X, $m^\alpha_X = X^\alpha$, we obtain

$$\tilde{\Box} = \tfrac{1}{2}g_{L\underline{L}}L^\alpha\underline{L}^\beta\partial^2_{\alpha\beta} + \tfrac{1}{4}g_{LL}\underline{L}^\alpha\underline{L}^\beta\partial^2_{\alpha\beta} - g_{aL}\underline{L}^\alpha e^\beta_a\partial^2_{\alpha\beta} + \gamma^{\alpha\beta}\partial^2_{\alpha\beta},$$

with

$$\gamma^{\alpha\beta}\partial^2_{\alpha\beta} = -g_{a\underline{L}}L^\alpha e^\beta_a \partial^2_{\alpha\beta} + \tfrac{1}{4}g_{\underline{L}\underline{L}}L^\alpha L^\beta \partial^2_{\alpha\beta} + g_{ab}e^\alpha_a e^\beta_b \partial^2_{\alpha\beta}.$$

In other words, $\gamma^{\alpha\beta}\partial^2_{\alpha\beta}$ is the part of $\tilde{\Box}$ which is expressed with *two* good derivatives. Setting

$$L_1 = -\tfrac{1}{2}g_{L\underline{L}}L - \tfrac{1}{2}g_{LL}\underline{L} + g_{aL}e_a,$$

we obtain

$$\tilde{\Box} = -L_1^\alpha \underline{L}^\beta \partial^2_{\alpha\beta} + \gamma^{\alpha\beta}\partial^2_{\alpha\beta}.$$

2. A modified optical function

a. After some rough estimates, it turns out that one can discard a part of L_1 and define

$$L_2 = L - \tfrac{1}{4}g_{LL}\underline{L}$$

as a substitute for the standard L in the transport equations. One also introduces, as a substitute for the optical function $u = r - t$, the **modified optical function** ρ defined by

$$|r-t| \geq t/2 \Rightarrow \rho(x,t) = r - t,\ |r-t| \leq t/2 \Rightarrow L_2\rho = 0.$$

Introducing the coordinates $q = r - t,\ p = r + t$, the key step of the proof is to obtain, by integration along L_2, for some $0 < \nu < 1$, the estimates

$$|\nabla\phi| \leq C\epsilon(1+t)^{-1}(1+|\rho|)^{-\nu},\ |\nabla^2\phi| \leq C\epsilon(1+t)^{-1}(1+|\rho|)^{-1-\nu}|\partial_q\rho|.$$

Note that these estimates are similar in form to what would follow from the Klainerman inequality, ρ replacing $r - t$. The point here is that $\nabla^2\phi$ behaves almost like a free solution.

b. Once these estimates are obtained, the rest of the proof is by commuting products Z^k to the equation, a procedure which requires serveral delicate arguments. As a result, one obtains, for some constant C,

$$E_N(t) \leq 8N\epsilon^2(1+t)^{C\epsilon}.$$

Taking $C\epsilon_0 \leq \delta$ finishes the proof by induction.

One may wonder whether the energy E_N actually grows with t or if this is just an artefact of the proof. We believe, by analogy with similar problems where such a growth has been proved, that this is a true phenomenon.

9.3 Low regularity well-posedness for quasilinear wave equations

Now we discuss the work of Klainerman and Rodnianski [31], which we used in chapter 7. The problem is the well-posedness of the (local) Cauchy problem with nonsmooth data

$$-\partial_t^2 v + g^{ij}(v)\partial_{ij}^2 v = N(v, \partial v),\ v(x,0) = v_0(x),\ (\partial_t v)(x,0) = v_1(x).$$

We assume $(v_0, v_1) \in H^s \times H^{s-1}$, and N quadratic in ∂v. The result of [31] (under some technical assumptions that we skip) is the following.

Theorem *The Cauchy problem has a unique local solution for* $s > s_c = 2 + \frac{2-\sqrt{3}}{2}$.

Let us recall that standard methods give the well-posedness of this Cauchy problem for $s > \frac{5}{2}$ (see [42] for instance), and that $\frac{2-\sqrt{3}}{2} \sim 0.13$. Though better results have been proved on this problem ([35], [45]), we discuss [31] as an instructive example. In contrast with the first two examples above, this is an example where one does not use the standard Lorentz fields Z, but develops the specific geometry of the problem.

1. Some words about the Littlewood–Paley theory To understand the context of the proof of the theorem, one has to know some definitions and ideas from the Littlewood–Paley theory (see [10] for details).

Let $\psi \in C_0^\infty(\mathbf{R}^n)$, $\psi(\xi) = 1$ for $|\xi| \le \frac{1}{2}$, $\psi(\xi) = 0$ for $|\xi| \ge 1$. Set

$$\phi(\xi) = \psi\left(\frac{\xi}{2}\right) - \psi(\xi),$$

so that ψ is supported in a ball while ϕ is supported in the shell $\frac{1}{2} \le |\xi| \le 2$. Moreover, for all ξ,

$$1 = \psi(\xi) + \sum_{p \ge 0} \phi(2^{-p}\xi).$$

We have constructed a dyadic partition of unity (in which there are at most two nonzero terms). For any $v \in \mathcal{S}'(\mathbf{R}^n)$, we set

$$v_{-1} = S_0 v = \psi(D)v,\ v_p = \phi(2^{-p}D)v,$$

where the notation $w = \psi(D)v$ means for the Fourier transforms $\hat{w}(\xi) = \psi(\xi)\hat{v}(\xi)$. In this way, we decompose v into

$$v = S_0 v + \sum_{p \ge 0} v_p = \sum_{p \ge -1} v_p,$$

each term having its Fourier transform compactly supported. Each term is, of course, a C^∞ function, and the regularity of v is read from the decay properties of the v_p.

In particular, a smoothing operator is defined by

$$S_p(D)v = \sum_{-1 \le q \le p-1} v_q.$$

This is nothing other than the Fourier truncation by $\psi(2^{-p}\xi)$.

2. The Strichartz inequality

Theorem *Let (p, q) be such that*

$$p > 2, \ \frac{1}{p} + \frac{3}{q} = \frac{1}{2}.$$

There exists a constant $C > 0$ such that, for all $T > 0$, all

$$v_0 \in H^1(\mathbf{R}^3), \ v_1 \in L^2(\mathbf{R}^3), \ f \in L^1([0, T], L^2(\mathbf{R}^3)),$$

the solution v of the Cauchy problem

$$\Box v = f, \ v(x, 0) = v_0(x), \ (\partial_t v)(x, 0) = v_1(x)$$

satisfies the inequality

$$||v||_{L^p([0,T],L^q(\mathbf{R}^3))} \le C \left\{ ||\nabla v_0||_{L^2(\mathbf{R}^3)} + ||v_1||_{L^2(\mathbf{R}^3)} + \int_0^T ||f(\cdot, t)||_{L^2(\mathbf{R}^3)} dt \right\}.$$

We do not want to prove this estimate completely here. We point out, however, that it is a consequence, using standard procedures (see [28] for instance), of the following *decay* inequality.

Lemma *Let v be the solution of the Cauchy problem*

$$\Box v = 0, \ v(x, 0) = v_0(x), \ (\partial_t v)(x, 0) = v_1(x),$$

where v_0 and v_1 are assumed to have their Fourier transforms supported in a shell $\frac{1}{2} \le |\xi| \le 2$. Then

$$||v(\cdot, t)||_{L^\infty} \le C t^{-1} (||\nabla^2 v_0||_{L^1} + ||\nabla v_1||_{L^1}).$$

Note that the presence of derivatives of v_0 and v_1 is irrelevant in the inequality, because of the assumption on $\hat{v}_i$; however, they make the inequality scale invariant, meaning that it does not change if we replace v by v_λ:

$$v_\lambda(x, t) = v(\lambda x, \lambda t).$$

Hence, to prove the Strichartz inequality, we can use the Klainerman inequality to prove the required decay estimate above.

3. Decay estimate and Lorentz fields Let us sketch the proof of the above decay lemma. We can assume $v_1 = 0$. Let us cover $\mathbf{R}^3$ by a union of discs D_I centered at the point $I \in \mathbf{Z}^3$ such that each D_I intersects at most a fixed finite number of discs D_J. Consider a C^∞ partition of unity χ_I supported in D_I. We can arrange

$$\sum_{I \in \mathbf{Z}^3} |\nabla^j \chi_I| \leq C_j.$$

Now, we localize the data v_0, $v_0 = \sum v_I^0$, $v_I^0 = \chi_I v_0$, and consider the function v_I solution of the wave equation with data $(v_I^0, 0)$. Of course, $v = \sum v_I$. Suppose that we have, for each I,

$$|\nabla v_I|(x,t) \leq C(1+t)^{-1} \sum_{j \leq 5} ||\nabla^j v_I^0||_{L^1}.$$

Summing over I, we obtain as desired

$$|\nabla v| \leq C(1+t)^{-1} ||v_0||_{L^1} \sum_{j \leq 5} ||\nabla^j \chi_I||_{L^\infty} \leq C(1+t)^{-1} ||v_0||_{L^1}.$$

Here, we have used the fact that $||\nabla^j v_0||_{L^1} \leq C ||v_0||_{L^1}$.

To prove the inequality on v_I, it is enough to do it for one of them, say $I = 0$. Then the Klainerman inequality and the energy inequality give us

$$\begin{aligned} |\nabla v_I|(x,t) &\leq C(1+t)^{-1} \sum_{k \leq 2} ||(\nabla Z^k v_I)(\cdot, t)||_{L^2} \\ &\leq C(1+t)^{-1} \sum_{k \leq 2} ||(\nabla Z^k v_I)(\cdot, 0)||_{L^2}. \end{aligned}$$

Since v_I^0 is supported in a fixed ball,

$$\sum_{k \leq 2} ||(\nabla Z^k v_I(\cdot, 0)||_{L^2} \leq C \sum_{k \leq 3} ||\nabla^k v_I^0||_{L^2}.$$

From the Sobolev inequality

$$||w||_{L^2} \leq C \sum_{j \leq 2} ||\partial^j w||_{L^1},$$

we finally get

$$|\nabla v_I|(x,t) \leq C(1+t)^{-1} \sum_{j \leq 5} ||\nabla^j v_I^0||_{L^1}.$$

This approach of the Strichartz inequality is likely also to work for variable coefficient wave equations.

4. Sketch of the proof of the well-posedness theorem The first step of the proof, which is due to Bahouri and Chemin [11], [12], is essentially to reduce the problem to proving a Strichartz estimate for some variable coefficient wave equation. This *linear* wave equation is associated to a (split) metric h_λ, depending on some parameter λ, which is a smoothed rescaled version of the original metric g. The precise behavior (with respect to λ) of h_λ and its derivatives reflects the smoothness assumptions on g.

In the original paper [11], the Strichartz estimate was proved using parametrices; in the "vector field approach " of [28], the Strichartz estimate follows from a decay inequality which is obtained in a way analogous to what we have done in the simple case of $\Box$ above.

The strategy of [31] is as follows. First, the authors define a "canonical" optical function u as being t on the t-axis and "having forward light cones with vertices on the time axis" as level surfaces. This refers to the construction explained in section 3.3. There seems to be no special reason for this choice, except its natural character; the normalization on the time axis reflects the will of imitating the flat case. Then, the full machinery of chapter 7 is developed, along with the use of conformal inequalities as described in chapter 4. We refer to [31] for the actual implementation of these techniques.

9.4 Stability of Minkowski spacetime (first version)

We refer here to the book by Klainerman and Nicolò *The Evolution Problem in General Relativity* [29], the previous book by Christodoulou and Klainerman [18] being more difficult to access. The goal is to solve the Einstein equations with initial data close to the flat Minkowski metric. More precisely, we look for a metric g, close to the Minkowski metric, for which the Ricci tensor R is identically zero (these are the simplest Einstein vacuum equations), and which extends for all t the Cauchy data given on $\{t = 0\}$ (in a sense which has to be made precise).

There are many ideas in this long work, but two of them seem especially relevant in the context of this book:

(i) *The authors never use the time variable. In the proof, they construct two optical functions u and $\underline{u}$ (substitutes for the usual $u = t - r$ and $\underline{u} = r + t$ of the flat case), and use the frame associated to these two functions.*

(ii) *The authors do not deal with the wave equation $\Box$, but with the Bianchi equations.*

We postponed the discussion of this case until now because of its complexity, though it presents some features analogous to the case of Maxwell equations.

1. Bianchi equations and energy inequalities The Bianchi equations (or second Bianchi identity) are first order equations on the curvature tensor R

$$D_{[\lambda} R_{\gamma\delta]\alpha\beta} = 0.$$

The bracket here means that we take the sum for the circular permutation of the indices. The idea is, of course, that these equations directly control the second order derivatives of g through R. One then has a better chance to recover an induction hypothesis on g without losing derivatives. The drawback is that these equations are more difficult to handle than a wave equation, or even the Maxwell system. The general character of this strategy, which Christodoulou calls the "first method," is sketched in the prolog of [16].

a. If the metric g satisfies the vacuum Einstein equations $R_{\mu\nu} = 0$, the curvature tensor R is traceless, by definition. Since we want to handle the Bianchi equations by the method of energy inequalities and commuting fields, we have to define a concept containing the curvature tensor R and some of its Lie derivatives. Hence, generally, we define a Weyl field as a traceless 4-tensor W, that is, satisfying

$$g^{\alpha\gamma} W_{\alpha\beta\gamma\delta} = 0$$

with the symmetries of the curvature tensor

$$W_{\alpha\beta\gamma\delta} = -W_{\beta\alpha\gamma\delta} = -W_{\alpha\beta\delta\gamma} = W_{\gamma\delta\alpha\beta},$$
$$W_{\alpha\beta\gamma\delta} + W_{\alpha\gamma\delta\beta} + W_{\alpha\delta\beta\gamma} = 0.$$

The tensor W is said to satisfy the Bianchi equations if

$$D_{[\lambda} W_{\gamma\delta]\alpha\beta} = 0.$$

b. Just as we did for Maxwell equations, one can define a dual tensor $*W$ by

$$*W_{\alpha\beta\gamma\delta} = \tfrac{1}{2}\epsilon_{\alpha\beta\mu\nu} W^{\mu\nu}{}_{\gamma\delta}.$$

Here ϵ is the volume form, and $*(*W) = -W$ as in the case of Maxwell equations. As in chapter 4, there exists an energy machinery to prove energy inequalities for the Bianchi equations that we explain here without proof. We define the energy–momentum tensor (called here the Bel–Robinson tensor):

$$Q_{\alpha\beta\gamma\delta} = W_{\alpha\rho\gamma\sigma} W_{\beta\ \delta}^{\ \rho\ \sigma} + *W_{\alpha\rho\gamma\sigma} * W_{\beta\ \delta}^{\ \rho\ \sigma}.$$

If W is a Weyl field satisfying the Bianchi equations, then

$$D^\alpha Q_{\alpha\beta\gamma\delta} = 0.$$

This is, of course, similar to the formula proved in chapter 4. Moreover, Q has the *positivity property*: if X, Y, Z, T are nonspacelike future oriented fields,

$$Q(X, Y, Z, T) \geq 0.$$

To prove an energy inequality, one chooses *three* multipliers X, Y, Z and sets

$$P_\alpha = Q_{\alpha\beta\gamma\delta} X^\beta Y^\gamma Z^\delta.$$

For a solution W of the Bianchi equations, we have then

$$\operatorname{div} P = \tfrac{1}{2} Q_{\alpha\beta\gamma\delta}[{}^{(X)}\pi^{\alpha\beta} Y^\gamma Z^\delta + {}^{(Y)}\pi^{\alpha\gamma} X^\beta Z^\delta + {}^{(Z)}\pi^{\alpha\delta} X^\beta Y^\gamma].$$

This is the analog of the key formula of section 4.3. We see that we now have the choice of three multipliers, which gives a lot of flexibility. In [29], many choices appear in which the multipliers are picked from the fields $T_0 = \frac{1}{2}(L + \underline{L})$ (standard choice), $K_0 = \frac{1}{2}(u^2 \underline{L} + \underline{u}^2 L)$ (conformal choice), L, or $\underline{L}$.

2. Induction We first have to notice that the authors do not proceed by induction on time, since there is no time! To describe the induction process used in [29], let us assume that we are dealing with the two standard optical functions $u = r - t$, $\underline{u} = t + r$. Consider the exterior region

$$Ext = \{(x, t),\ u(x, t) \geq M\}.$$

This exterior region is foliated by the level surfaces of $\underline{u}$. The induction is on $\underline{u}$: assuming some induction hypothesis $\mathcal{P}_{\underline{u}^0}$ in the region $Ext \cap \{\underline{u} \leq \underline{u}^0\}$, one proves that the solution extends to the bigger region $Ext \cap \{\underline{u} \leq \underline{u}^0 + \epsilon\}$ and satisfies $\mathcal{P}_{\underline{u}^0 + \epsilon}$ there. Note that the extension of the solution is done across the characteristic surface $\{\underline{u} = \underline{u}^0\}$. This description, of course, has to be adapted to the actual optical functions which we now describe.

3. Optical functions Consider a bounded region K of spacetime whose boundary is formed by

(i) a portion of the spacelike initial hypersurface Σ_0,
(ii) a portion of the null outgoing hypersurface C_0,
(iii) a portion of the null incoming hypersurface $\underline{C}_*$.

We want to construct two optical functions u and $\underline{u}$ such that C_0 is contained in a level set of u, and $\underline{C}_*$ is contained in a level set of $\underline{u}$. In fact, in the approach of [29], the authors proceed by induction on the "last slice" $\underline{C}_*$, instead of proceeding by induction on time as in example 9.1 for instance.

a. To construct $\underline{u}$, we prescribe it as w on the initial hypersurface Σ_0. The function w is chosen so that its level sets form a sphere foliation of Σ_0 with properties that we explain now. The natural frame associated with such a foliation is (N, e_a), where

$$N = |\nabla w|^{-1} \nabla w$$

is the unit vector normal to the leaves of the foliation, and (e_a) form an orthonormal basis of the two-dimensional leaves. If θ denotes the second form of the foliation (in Σ_0), one can establish the equation (we keep the notation of [29] for convenience)

$$N(\operatorname{tr}\theta) + \tfrac{1}{2}(\operatorname{tr}\theta)^2 = -(\not{\Delta} \log a + \rho) + [-|\not{\nabla} \log a|^2 - |\hat{\theta}|^2 + g(k)],$$

where

$$a = |\nabla w|^{-1}, \ \rho = -\tfrac{1}{4} R_{3434}, \ g(k) = k_{NN}^2 + \sum |k_{Na}|^2.$$

Here, $\not{\Delta}$ and $\not{\nabla}$ refer to the induced connexion on the sphere foliation, and the frame implicitly used in the notation R_{3434} is

$$e_3 = N - T_0, \ e_4 = N + T_0,$$

where T_0 is a unit vector orthogonal to Σ_0. In order to save derivatives of θ, we want w to be constant on the trace of $\underline{C}_*$ on Σ_0, and a to satisfy the elliptic equation on the leaves:

$$\not{\Delta} \log a = -(\rho - \bar{\rho}), \ \overline{\log a} = 0,$$

where the overbar denotes the mean value on the leaves. The existence of such a function w, of course, requires a proof, and we refer the reader to [29] for details. The point we want to make here is that $\underline{u}$ is constructed in a very careful way, in accordance with smoothness requirements.

b. The optical function u is the outgoing solution of the eikonal equation with initial condition $u = u_*$ on the last slice $\underline{C}_*$. The function u_* is the solution of a highly nontrivial system which we do not discuss here.

The careful construction of both optical functions u and $\underline{u}$ and their associated frame makes it possible to obtain specific decay properties for the various components of R on this frame. These are called "peeling properties" (see [30]).

4. Commutations In [29], the authors also need to commute vector fields with the Bianchi equations. We already know from the case of Maxwell equations that the good way to do this is to consider $\mathcal{L}_X W$. There is, however, some technical difficulty: namely, this Lie derivative of W is no longer a Weyl field! This forces us to define a modified Lie derivative $\hat{\mathcal{L}}_X W$, which is $\mathcal{L}_X W$ plus some linear combination of components of W. We will not go any further in this direction, and refer the reader to [29]. What are the modified Lorentz fields that nearly commute with the Bianchi equations? They are constructed from the sphere foliation as explained in section 6.3.

9.5 L^2 conjecture on the curvature

We refer the reader here to two series of works:

(i) the papers [33], [35], [36], where the local well-posedness of the vacuum Einstein equations is proved with an initial curvature in H^{+0},
(ii) the papers [32], [34], [37] starting the proof of the same result with curvature only in L^2.

The general framework is very similar to that of the example 9.4. The challenge is to control the geometry of null geodesic cones, and of the associated optical functions and frames, using only L^2 bounds on some components of the curvature. The control of this geometry will allow us to use the machinery of chapter 7 to obtain estimates of the curvature via the Bianchi equations, as explained in the preceding section. We sketch here the issue of the boundedness of $\operatorname{tr}\chi$, the importance of which we first explain.

1. tr χ and the geometry Let u be a given optical function. Let S_0 be a fixed "2-sphere" in an initial spacelike hypersurface Σ_0. Let u be constant on S_0 and consider the hypersurface H which is the union of the integral curves of $L = -\nabla u$ starting from S_0. Then u is constant on H, and H is a null hypersurface. Let s be the function on H defined by

$$Ls = 1, \; m \in S_0 \Rightarrow s(m) = 0.$$

The image of S_0 by the flow of L at time s_0 is the level surface S_{s_0} of s, and the "2-spheres" S_s form a foliation, called the "geodesic foliation" of H. The null frame on H we are working with is associated to this sphere foliation as explained in chapter 2.

We pick up on S_0 coordinates $\omega = (\omega^1, \omega^2)$ and define coordinates on H following the trajectories of L; more precisely, if m is the image of the point of S_0 of coordinates ω by the flow of L at time s, the coordinates of m are

(s, ω). The importance of $\operatorname{tr}\chi$ with respect to the foliation S_s is displayed in the following theorem.

Theorem *Let dA_s be the area element on S_s, and $|S_s| = \int_{S_s} dA_s$ be the area of S_s. Then*

$$\frac{d}{ds}|S_s| = \int_{S_s} (\operatorname{tr}\chi) dA_s.$$

Proof We first prove that, with the coordinates we introduced on H,

$$\partial_s(g_{ab}) = 2\chi_{ab}.$$

This is due to the fact that $L = \partial_s$:

$$\begin{aligned} L\langle\partial_a, \partial_b\rangle &= \langle D_L\partial_a, \partial_b\rangle + \langle\partial_a, D_L\partial_b\rangle \\ &= \langle D_a L, \partial_b\rangle + \langle\partial_a, D_b L\rangle = 2\chi_{ab}, \end{aligned}$$

since $[L, \partial_a] = 0$. Denoting by γ the restriction of g to the spheres, this implies

$$\partial_s(|\gamma|^{\frac{1}{2}}) = |\gamma|^{\frac{1}{2}} \operatorname{tr}\chi.$$

Now

$$|S_s| = \int |\gamma|^{\frac{1}{2}} d\omega^1 d\omega^2,$$

$$(d/ds)|S_s| = \int tr\,\chi|\gamma|^{\frac{1}{2}} d\omega^1 d\omega^2 = \int_{S_s} \operatorname{tr}\chi dA_s. \qquad \diamond$$

2. Control of χ and curvature

a. Define some components of the curvature tensor R by

$$\beta_a = R_{La\underline{L}L},\ \rho = \tfrac{1}{4}R_{\underline{L}L\underline{L}L},\ \sigma = *R_{\underline{L}L\underline{L}L}.$$

Assume that these components are bounded in $L^2(H)$ by R_0. We want to use the machinery of chapter 7 to obtain a bound for $\operatorname{tr}\chi$ in L^∞, carefully counting derivatives.

To do this, we come back to the transport equation for $\operatorname{tr}\chi$, which is here, since the Ricci tensor is zero,

$$L(\operatorname{tr}\chi) + \tfrac{1}{2}(\operatorname{tr}\chi)^2 = -|\hat{\chi}|^2.$$

To obtain a L^∞ control of $\operatorname{tr}\chi$, we need to control $\int_\Gamma |\hat{\chi}|^2$ on each of the geodesics Γ which foliate H. We turn then to the Codazzi equation on $\hat{\chi}$, which is here

$$\operatorname{div}\hat{\chi} = -\beta + \tfrac{1}{2}\,\nabla\!\!\!/\operatorname{tr}\chi + \cdots,$$

where the dots denote terms which cause no problems. Denoting by D^{-1} the pseudodifferential operator of order -1 which solves the elliptic system on $\hat{\chi}$, we have to bound

$$I_1 = \int_\Gamma |D^{-1}\beta|^2, \; I_2 = \int_\Gamma |D^{-1} \not\nabla \mathrm{tr}\, \chi|^2.$$

b. The bound on β implies that $D^{-1}\beta$ is bounded in $H^s(H)$ for $s = 1$, and we cannot consider its trace on the curve Γ of codimension 2, since this would require $s > \frac{2}{2} = 1$. In a way which is analogous to what was done with the special component R_{44} in section 7.3, we investigate now the *special structure* of β. To do this, we write the Bianchi equations in our frame, and obtain

$$\mathrm{div}\, \beta = D_L \rho + \cdots, \; \mathrm{curl}\, \beta = -D_L \sigma + \cdots,$$

where as usual the dots represent harmless terms. In short, we write the solution of this elliptic system $\beta = D^{-1}(L(\rho), L(\sigma))$, thus obtaining

$$D^{-1}\beta = \nabla_L Q + \cdots, \; Q = D^{-2}(\rho, \sigma).$$

The integral I_1 is bounded by $||Q|_\Gamma||^2_{H^1}$, which is itself bounded by

$$||Q||^2_{H^2(H)} \le C||(\rho, \sigma)||^2_{L^2(H)} \le C R_0^2.$$

c. To bound $||\mathrm{tr}\, \chi||_{L^\infty}$ using the transport equation, we also have to bound I_2 with this same norm. The difficulty here is that $D^{-1} \not\nabla$ is a zero order pseudodifferential operator on the spheres, which does not act on L^∞. This forces the authors to work in Besov spaces, and to construct a Littlewood–Paley theory adapted to low regularity submanifolds. This leads to considerable developments which are beyond the scope of this introduction. A related approach that is still being devoloped, based on a Lax type parametrix construction, was discussed by Szeftel in his Cours Peccot at the Collège de France (2007).

9.6 Stability of Minkowski spacetime (second version)

For quite a long time, it was believed that working in harmonic coordinates for the Einstein equations could only lead to local (in time) existence results, see, for instance, the work of Choquet-Bruhat [13]. To prove global results the idea was then to avoid coordinates altogether, as was the case for the first version of the proof mentioned in section 9.4. In this second version [41], however, just as in the section 9.2, it turns out that one can use the standard Lorentz fields to handle the problem of small solutions.

1. Harmonic coordinates Let us explain first the use of "harmonic coordinates." When we use harmonic coordinates we are working on $\mathbf{R}^4$ with coordinates x^α, each one of them being a solution of the wave equation $\Box_g x^\alpha = 0$. From the formula for $\Box$ in section 3.3.2, this means, for each μ,

$$\partial^\alpha g_{\alpha\mu} = \tfrac{1}{2} g^{\alpha\beta} \partial_\mu g_{\alpha\beta}.$$

It also means that the lower order terms in $\Box$ are identically zero:

$$\Box \equiv g^{\alpha\beta} \partial^2_{\alpha\beta}.$$

If we take the ν derivative of the formula $\partial^\alpha g_{\alpha\mu} = \frac{1}{2} g^{\alpha\beta} \partial_\mu g_{\alpha\beta}$, we get

$$\partial^\alpha \partial_\nu g_{\alpha\mu} = \tfrac{1}{2} g^{\alpha\beta} \partial^2_{\mu\nu} g_{\alpha\beta} + q_{\mu\nu},$$

where $q_{\mu\nu}$ is a quadratic expression in the first order derivatives of g:

$$q_{\mu\nu} = \tfrac{1}{2} \partial_\nu (g^{\alpha\beta}) \partial_\mu g_{\alpha\beta} - (\partial_\nu g^{\alpha\beta}) \partial_\beta g_{\alpha\mu}.$$

Exchanging μ and ν and summing, we obtain

$$\partial^\alpha \partial_\nu g_{\alpha\mu} + \partial^\alpha \partial_\mu g_{\alpha\nu} - g^{\alpha\beta} \partial^2_{\mu\nu} g_{\alpha\beta} = q_{\mu\nu} + q_{\nu\mu}.$$

Using the explicit formula from section 7.1 for the Ricci tensor, we observe that, in harmonic coordinates, the first three terms are the same as the left-hand side of the equality above. Hence the vacuum Einstein equations can be written

$$\Box g_{\mu\nu} = F_{\mu\nu}(g)(\partial g, \partial g),$$

for some appropriate expressions $F_{\mu\nu}$, quadratic in ∂g. This is the only known way to display the *hyperbolic character* of Einstein equations.

Besides reducing Einstein equations to a hyperbolic system, the point of harmonic coordinates is this: *suppose g satisfies the system*

$$g^{\alpha\beta} \partial^2_{\alpha\beta} g_{\mu\nu} = F_{\mu\nu}(g)(\partial g, \partial g)$$

with initial values $(g, \partial_t g) = (g_0, g_1)$ *on* $\{t = 0\}$ *satisfying the harmonic coordinates relations; then the harmonic coordinates relations are true for all* t, *and* g *is, in fact, a solution of Einstein equations.*

We refer the reader to [29] for a proof of this well-known fact.

2. Good derivatives and good components

a. From the form of the Einsten equations obtained in **1.**, we see that we have to work with a *diagonal* (in its principal part) system of wave equations, *coupled* with the first order conditions of harmonic coordinates. In the *scalar* case in section 9.2, we concentrated on the good derivatives of the solution ϕ; since we work here with a system on the tensor g, we will concentrate not only on the good derivatives of single components $g_{\alpha\beta}$, but also on derivatives of good components of g. The harmonic coordinates relation provides precisely a *link between good derivatives and good components* of the metric. With the notation of [41], let T be one good derivative L, e_1, e_2. Defining the perturbation h of the Minkowski metric by $g^{\alpha\beta} = m^{\alpha\beta} + h^{\alpha\beta}$, we have

$$T^\mu \partial^\alpha h_{\alpha\mu} = T(h^\alpha_\alpha) + O(h\partial h).$$

On the left-hand side, we have for fixed μ, modulo $O(h\partial h)$ terms, the trace of the tensor

$$(X, Y) \mapsto D_X h(Y, \partial_\mu).$$

In the null frame $(e_1, e_2, \underline{L}, L)$, this trace is

$$D_a h(e_a, \partial_\mu) - \tfrac{1}{2} D_L h(\underline{L}, \partial_\mu) - \tfrac{1}{2} D_{\underline{L}} h(L, \partial_\mu).$$

This implies the relation

$$|(\partial h)_{LT}| \leq |Th| + O(h\partial h).$$

In words this can be expressed: the good component LT of any derivative of h is controlled, modulo harmless terms, by all components of a good derivative of h. This is the *"duality"* specific of this system.

b. As in section 9.2, we start with the induction hypothesis

$$E_N(t) \equiv \frac{1}{2} \sum_{k\leq N} \int |(\partial Z^k \phi)(x,t)|^2 dx \leq 64\epsilon^2(1+t)^{2\delta}$$

for some $0 < \delta < 1$. The above estimate and other similar estimates following the same duality principle and using the harmonic coordinates condition give, for some $\gamma > 0$,

$$|(\partial h)_{LT}| + |(\partial Z h)_{LL}| \leq C\epsilon(1+t+r)^{-1-2\gamma},$$

$$|h_{LT}| + |(Zh)_{LL}| \leq C\epsilon(1+t+r)^{-1}\langle r-t\rangle.$$

3. Improved standard energy inequality In [41], the authors establish the following improved energy inequality, which is very close to the one in section 5.3.

Theorem *Assume that the metric g satisfies the decay estimates*

$$\langle r-t\rangle^{-1}|h_{LL}| + |(\partial h)_{LL}| + |Th| \leq C\epsilon(1+t)^{-1},$$

$$\langle r-t\rangle^{-1}|h| + |\partial h| \leq C\epsilon(1+t)^{-\frac{1}{2}}(\langle r-t\rangle^{-\frac{1}{2}-\gamma}.$$

Then, for any $0 < \gamma \leq \frac{1}{2}$*, there exists* $\epsilon_0 > 0$ *such that, for* $\epsilon \leq \epsilon_0$*,*

$$E_\phi(T) + \tfrac{1}{2}\gamma \int_{0\leq t\leq T} \langle r - t\rangle^{-1-2\gamma} \sum |T\phi|^2 dxdt$$
$$\leq 8E_\phi(0) + C\epsilon \int_{0\leq t\leq T} (1+t)^{-1}|\partial\phi|^2 dxdt + 16 \int_{0\leq t\leq T} |\Box\phi||\partial_t\phi| dxdt.$$

Note that this inequality gives the classical improved energy inequality with an amplification factor $(1+t)^{C\epsilon}$. The estimates already proved on h fit with the assumptions of the theorem, and this is one of the keys to the proof.

4. The source terms and the weak null condition

a. ***Asymptotic analysis*** Let us consider again the Cauchy problem with small data for a general quasilinear wave equation

$$\Box\phi + \sum g^{\alpha\beta\gamma}(\partial_\gamma\phi)(\partial^2_{\alpha\beta}\phi) = 0,\ \phi(x,0) = \epsilon\phi_0(x),\ \partial_t\phi(x,0) = \epsilon\phi_1(x).$$

For t not too large, ϕ is small and the quadratic terms are negligible, so that ϕ behaves essentially like ϵ times ϕ^1, the solution of the Cauchy problem

$$\Box\phi^1 = 0,\ \phi^1(x,0) = \phi_0(x),\ (\partial_t\phi^1)(x,0) = \phi_1(x).$$

To make this intuition more precise, formally let

$$\phi = \epsilon\phi^1 + \epsilon^2\phi^2 + \cdots.$$

We find that

$$\Box\phi^2 + \sum g^{\alpha\beta\gamma}(\partial_\gamma\phi^1)(\partial^2_{\alpha\beta}\phi^1) = 0.$$

We know that (with $\sigma = r - t$)

$$\phi^1 = r^{-1}F^1(r-t,\omega,r^{-1}) = r^{-1}F^1_0(r-t,\omega) + O(r^{-2}),$$
$$F^1_0(\sigma,\omega) = F^1(\sigma,\omega,0)$$

for some C^∞ function F^1 (see [9], [24] for details). The nonlinear terms in ϕ^1 in the above equation are

$$r^{-2}g(\omega)(\partial_\sigma F^1_0)(\partial^2_\sigma F^1_0) + O(r^{-3}),\ g(\omega) = \sum g^{\alpha\beta\gamma}\omega_\alpha\omega_\beta\omega_\gamma.$$

We deduce from this that $\phi^2 \sim (g\log t/4r)(\partial_\sigma F^1_0)^2$, so that

$$\phi = \epsilon r^{-1}[F^1_0 + \frac{g}{4}(\epsilon\log t)(\partial_\sigma F^1_0)^2 + \cdots].$$

Hence the effect of the nonlinear terms of the equation is felt through a slow time correction, the slow time being the variable $\tau = \epsilon\log t$. This suggests trying for ϕ the ansatz

$$\phi \sim \epsilon r^{-1} G(r-t, \omega, \tau).$$

Since

$$\Box r^{-1} G = \frac{-2}{rt} \partial^2_{\sigma\tau} G + O(r^{-3}),$$

the function G should satisfy the equation

$$-2\partial^2_{\sigma\tau} G + g(\omega)(\partial_\sigma G)(\partial^2_\sigma G) = 0.$$

This asymptotic analysis is due to Hörmander [24]. It is the starting point for the heuristic understanding of the lifespan of the solution: in the general case, the equation for G is nonlinear, so that G is expected to blowup in finite time τ_0, which suggests a lifespan $T_\epsilon \sim \exp(\tau_0/\epsilon)$ (compare with the theorem in section 9.4); if the equation satisfies the null condition, $g(\omega) \equiv 0$, suggesting global existence.

b. The weak null condition Suppose now we deal with a diagonal system of wave equations with source terms

$$\Box \phi^i = F^{\alpha\beta}_{ijk} (\partial_\alpha \phi^j)(\partial_\beta \phi^k), \ 1 \le i \le N,$$

where the coefficients $F^{\alpha\beta}_{ijk}$ are real constants. Using the same ansatz as before, we obtain the system

$$-2\partial^2_{\sigma\tau} G^i = \bar{F}_{ijk}(\partial_\sigma G^j)(\partial_\sigma G^k), \ \bar{F}_{ijk} = F^{\alpha\beta}_{ijk} \omega_\alpha \omega_\beta.$$

This is a system of ODEs in the variable τ on the functions $\partial_\sigma G^i$. In [40], the authors introduce the following definition.

Definition The original PDE system is said to satisfy the weak null condition if the ODE system on G has, for all data, global solutions growing at most exponentially with τ.

Since $\exp C\tau = t^{C\epsilon}$, the weak null condition suggests global existence for the PDE system. However, this is far from being proved, as shown in [8].

c. The Einstein equations If we forget that the wave operator associated with the metric g is not the standard $\Box$, we can think of the system of the Einstein equations in harmonic coordinates as being of the form discussed in **b**. With this approximation, it is shown in [40] that the Einstein equations satisfy the weak null condition. This is an important fact in the proof in [41].

The conclusion is that, in the study of Einstein equations, and more generally in the study of systems of wave equations or hyperbolic symmetric systems with an unknown $u \in \mathbf{R}^N$, it is an essential step to understand the "duality" between the good derivatives of u and its good components. Note that in the present case, as well as in the case of the Bianchi equations briefly discussed in section 9.3, the *same* null frame is used to identify the good derivatives $(e_1(h), e_2(h), L(h))$ and the good components $h_{LT}, \partial h_{LT}$. In a more general situation, it could occur that one has to construct a null frame in the physical space to identify the good derivatives of u, and another frame (with which properties?) in $\mathbf{R}^N$ to capture the good components of u. An example of this is to be found in [8].

9.7 The formation of black holes

We refer the reader here to the monograph [16] by Christodoulou. We will not discuss here the heart of the book which is what the author calls the "short pulse method" or "third method" (the first two methods work with Bianchi identities and with null frames, as explained in section 9.4). We only want to point out the construction of the optical functions, which is very close to that of section 9.3. The author first constructs a timelike geodesic line Γ_0, and considers the outgoing future null geodesic cones with vertices on Γ_0. The optical function u is then taken to have these cones C_u as level surfaces. For $\underline{u}$, its level surfaces are assumed to be the past incoming geodesic cones $\underline{C}_{\underline{u}}$ with vertices on Γ_0. The exact values of u and $\underline{u}$ on Γ_0 depend only on two functions of one variable, which leaves much less flexibility than in section 9.4, where u and $\underline{u}$ depended on the choices of two functions of three variables. Despite this fact, it turns out that these choices of u and $\underline{u}$ are relevant, since in the end some sphere of the foliation

$$S_{u\underline{u}} = C_u \cap \underline{C}_{\underline{u}}$$

turns out to be the desired "trapped surface." We will not explain this term, let us only say that a trapped surface indicates a black hole singularity, as explained in [23] for instance.

References

[1] Alinhac, S., The null condition for quasilinear wave equations in two space dimensions II, *Amer. J. Math.* **123** (2000), 1–31.

[2] Alinhac, S., The null condition for quasilinear wave equations in two space dimensions I, *Invent. Math.* **145** (2001), 597–618.

[3] Alinhac, S., An example of blowup at infinity for a quasilinear wave equation, *Astérisque* **284** (2003), 1–91.

[4] Alinhac, S., Méthodes géométriques dans l'étude des équations d'Einstein, *Bourbaki Seminar* **934** (2003–2004), 1–17.

[5] Alinhac, S., Remarks on energy inequalities for wave and Maxwell equations on a curved background, *Math. Ann.* **329** (2004), 707–722.

[6] Alinhac, S., Free decay of solutions to wave equations on a curved background, *Bull. Soc. Math. France* **133** (2005), 419–458.

[7] Alinhac, S., On the Morawetz–Keel–Smith–Sogge inequality for the wave equation on a curved background, *Publ. Res. Inst. Math. Sc. Kyoto* **42** (2006), 705–720.

[8] Alinhac, S., Semilinear hyperbolic systems with blowup at infinity, *Indiana Univ. Math. J.* **55** (2006), 1209–1232.

[9] Alinhac, S., *Hyperbolic Partial Differential Equations, an Elementary Introduction*, Universitext, New York: Springer Verlag (2009).

[10] Alinhac, S. and Gérard, P., *Pseudo-differential Operators and the Nash–Moser Theorem*, Grad. Stud. Math. 82, American Mathematical Society (2007).

[11] Bahouri, H. and Chemin, J.-Y., Equations d'ondes quasilinéaires et estimations de Strichartz, *Amer. J. Math.* **121** (1999), 1337–1377.

[12] Bahouri, H. and Chemin, J.-Y., Equations d'ondes quasilinéaires et effet dispersif, *Int. Math. Res. Not.* **21** (1999), 1141–1178.

[13] Choquet-Bruhat, Y., Théorèmes d'existence pour certains systèmes d'équations aux dérivées partielles non linéaires, *Acta Math.* **88** (1952), 141–225.

[14] Christodoulou, D., Global solutions for nonlinear hyperbolic equations for small data, *Comm. Pure Appl. Math.* **39** (1986), 267–282.

[15] Christodoulou D., *The Action Principle and Partial Differential Equations*, Ann. Math. Studies 146. Princeton: Princeton University Press (2000).

[16] Christodoulou, D., *The Formation of Black Holes in General Relativity*, EMS Monographs in Math. Zurich: European Mathematics Society (2008).

[17] Christodoulou, D. and Klainerman, S., Asymptotic properties of linear field equations in Minkowski space, *Comm. Pure Appl. Math.* **43** (1990), 137–199.

[18] Christodoulou, D. and Klainerman, S., *The Global Nonlinear Stability of the Minkowski space*, Princeton Math. Series 41. Princeton: Princeton University Press (1993).

[19] Dafermos, M. and Rodnianski, I., The redshift effect and radiation decay on black hole spacetimes, preprint, (2007).

[20] Dafermos, M. and Rodnianski, I., Lectures on black holes and linear waves, arXiv 0811.0354, (2008).

[21] Friedrich, H., Smoothness at null infinity and the structure of initial data, in *The Einstein Equations and the Large Scale Behavior of Gravitational Fields.* Basel, Birkhäuser (2004).

[22] Gallot, S., Hulin, D. and Lafontaine, J., *Riemannian Geometry*, Springer Universitext. New York: Springer Verlag (1990).

[23] Hawking, S. W., and Ellis, G. F. R., *The Large Scale Structure of Spacetime*, Cambridge Mono. Math. Physics. Cambridge: Cambridge University Press (1973).

[24] Hörmander, L., *Lectures on Nonlinear Hyperbolic Differential Equations*, Math. Appl. 26. New York: Springer Verlag (1997).

[25] Klainerman, S., Uniform decay estimates and the Lorentz invariance of the classical wave equation, *Comm. Pure Appl. Math* **38** (1985), 321–332.

[26] Klainerman, S., The null condition and global existence to nonlinear wave equations, *Lect Appl. Math.* **23** (1986), 293–326.

[27] Klainerman, S., Remarks on the global Sobolev inequalities in Minkowski space, *Comm. Pure Appl. Math.* **40** (1987), 111–117.

[28] Klainerman, S., A commuting vector field approach to Strichartz type inequalities and applications to quasilinear wave equations, *Int. Math. Res. Not.* **5** (2001), 221–274.

[29] Klainerman, S. and Nicolò, F., *The Evolution Problem in General Relativity*, Progr. Math. Physics 25. Boston: Birkhäuser (2003).

[30] Klainerman, S. and Nicolò, F., Peeling properties of asymptotic solutions to the Einstein-vacuum equations, *Class. Quant. Gravity* **20** (2003), 3215–3257.

[31] Klainerman, S. and Rodnianski, I., Improved local well-posedness for quasilinear wave equations in dimension three, *Duke Math. J.* **117** (2003), 1–124.

[32] Klainerman, S. and Rodnianski, I., Sharp trace theorems for null hypersurfaces on Einstein metrics with finite curvature flux, *Geom. Funct. Anal.* **16** (2006), 164–229.

[33] Klainerman, S. and Rodnianski, I., Ricci defects of microlocalized Einstein metrics, *J. Hyp. Diff. Eq.* **1** (2004), 85–113.

[34] Klainerman, S. and Rodnianski, I., Causal geometry of Einstein-vacuum spacetimes with finite curvature flux, *Invent. Math.* **159** (2005), 437–529.

[35] Klainerman, S. and Rodnianski, I., The causal structure of microlocalized rough Einstein metrics, *Ann. Math.* **161** (2005), 1143–1193.

[36] Klainerman, S. and Rodnianski, I., Bilinear estimates on curved space-times, *J. Hyp. Diff. Eq.* **2** (2005), 279–291.

[37] Klainerman, S. and Rodnianski, I., A geometric approach to the Littlewood–Paley theory, *Geom. Funct. Anal.* **16** (2006), 126–163.

[38] Klainerman, S. and Sideris, T., On almost global existence for nonrelativistic wave equations in 3D, *Comm. Pure Appl. Math.* **XLIX** (1996), 307–321.

[39] Lindblad, H., Global solutions of quasilinear wave equations, *Amer. J. Math.* **130** (2008), 115–157.

[40] Lindblad, H. and Rodnianski, I., The weak null condition for Einstein's equations, *C. R. Acad. Sci. Paris, Ser. I* **336** (2003), 901–906.

[41] Lindblad, H. and Rodnianski, I., Global existence for the Einstein vacuum equations in wave coordinates, *Comm. Math. Physics* **256** (2005), 43–110.

[42] Majda, A. J., *Compressible Fluid Flow and Systems of Conservation Laws in several variables*, Appl. Math. Sc. 53. New-York: Springer Verlag (1984).

[43] Penrose, R. and Rindler, W., *Spinors and Space-Time*, vols. 1 and 2. Cambridge: Cambridge University Press (1986).

[44] Rendall, A. D., *Partial Differential Equations in General Relativity*, Oxford Grad. Texts Math. Oxford: Oxford University Press (2008).

[45] Smith, H. F. and Tataru, D., Sharp local well-posedness results for the nonlinear wave equation, *Ann. Maths* **162** (2005), 291–366.

[46] Spivak M., *Calculus on Manifolds*. New York: Benjamin (1965).

[47] Wald R., *General Relativity*. Chicago: University of Chicago Press (1984).

Index

amplification factor, 36, 50, 84
Ansatz, 112
area element, 107
asymptotic analysis, 112

Bel-Robinson tensor, 103
Bianchi equations, 103
bicharacteristic, 23
black hole, 39, 113
blowup, 112
blowup criterion, 79
boundary terms, 35

Christoffel symbols, 18
Codazzi equation, 71
commutation formula, 59
commutation lemma, 85
conformal energy, 4, 53
conformal energy inequality, 4, 53, 58
conformal Killing, 32
connexion, 17
curvature tensor, 66

d'Alembertian, 21
decay estimate, 100
deformation tensor, 31
div–curl system, 73
divergence, 19
dual, 10
dual basis, 11
dual form, 41
dual tensor, 103

eikonal equation, 12
Einstein equations, 102, 112
electric field, 41
elliptic systems, 71
ellipticity, 73
energy, 34, 49
energy inequality, 2, 29, 33, 92, 104
energy–momentum tensor, 29, 43, 103
exterior derivative, 41

frame coefficients, 24, 68
future oriented, 30

geodesic, 22
geodesic cone, 23
ghost weight, 47, 92
global existence, 94, 96
good commutation condition, 61
good components, 110
good derivatives, 2, 45
gradient, 5, 10
Gronwall lemma, 35

harmonic coordinates, 109
Hessian, 21
hyperbolic, 1
hyperbolic rotations, 61
hyperbolic symmetric systems, 81

improved energy inequality, 47, 110
incoming light cones, 5
indices, 8
induced connexion, 20
induced metric, 20
induction, 104
  on time, 79, 84, 95, 97
induction hypothesis, 78, 86, 95
interior terms, 35

Kerr metric, 10, 13
Killing field, 32
Klainerman inequality, 3, 57
Klainerman method, 57, 101

Lie derivative, 31, 63
lifespan, 87
Littlewood–Paley theory, 99, 108
local existence theory, 79
Lorentz fields, 2
Lorentzian metrics, 9
low regularity well-posedness, 99

magnetic field, 41
Maxwell equations, 41, 51, 63
metric, 5, 8
metric connexion, 17
Minkowski metric, 9
mixed transport–elliptic system, 75
modified Lie derivative, 106
modified Lorentz fields, 61
modified optical function, 98
Morawetz inequality, 38
multiplier, 29

nonstandard 2-spheres, 15
null, 30
null condition, 90
null frame, 4, 12

optical function, 5, 12, 24, 68, 104
outgoing light cones, 5

past oriented, 30
Penrose diagramm, 9
photon sphere, 39
Poincaré inequality, 40
pointwise decay, 58
Poisson bracket, 40
positive field, 38
positivity property, 30, 43, 104
principal symbol, 23

quasilinear wave equation, 80, 89, 96
quasiradial frame, 14

Ricci tensor, 66

scalar curvature, 66
Schwarzschild metric, 9, 13, 39
second form, 20
slow time, 111
spacelike, 30
sphere foliation, 15, 69
spherical harmonics, 40
stability of Minkowski space, 102, 108
Stokes formula, 21
Strichartz inequality, 100
submanifold, 20
symmetric systems, 80

tensor, 10, 18, 31, 71
timelike, 30
trace, 11, 19
transport equation, 68
trapped surface, 113

volume form, 11

wave equation, 6
weak null condition, 112
weighted inequality, 36, 48
Weyl field, 103